Topics in Biological Inorganic Chemistry

Volume 3

Springer

Berlin
Heidelberg
New York
Barcelona
Hong Kong
London
Milan
Paris
Singapore
Tokyo

Calcium Homeostasis

Editors: E. Carafoli · J. Krebs

With contributions by
I.D. Campbell, E. Carafoli, A.K. Downing,
N. Gabellini, D. Guerini, P.A. Handford, M. Ikura,
G. Inesi, H. Kawasaki, C.B. Klee, J. Krebs,
R.H. Kretsinger, S. Nakayama, Ch. Toyoshima,
R.J.P. Williams, K.L. Yap, T. Yuan, A. Zatti

Springer

Volume Editors:

Professor Dr. Ernesto Carafoli
University of Padova
Department of Biological Chemistry
Viale G. Colombo 3
35121 Padova
Italy

Priv.-Doz. Dr. Joachim Krebs
Institut für Biochemie
ETH-Zentrum
8092 Zürich
Switzerland

ISSN 1437-7993
ISBN 3-540-67175-7 Springer Verlag Berlin Heidelberg New York

Library of Congress Cataloging-in-Publication Data applied for

Die Deutsche Bibliothek – CIP-Einheitsaufnahme
Calcium homeostasis/ed.: E. Carafoli; J. Krebs. With contributions by I.D. Campbell... – Berlin; Heidelberg; New York; Barcelona; Hong Kong; London; Milan; Paris; Singapore, Tokyo: Springer, 2000
(Topics in biological inorganic chemistry; Vol. 3)
ISBN 3-540-67175-7

Coverdesign: Friedhelm Steinen-Broo, Pau/Spain; MEDIO, Berlin
Typesetting: Scientific Publishing Services (P) Ltd, Madras

SPIN: 10735241 31/3020 – 5 4 3 2 1 0 – Printed on acid-free paper

Editorial Board of the Series

Preface to Volume 3

"Ja, Kalzium, das ist alles..."

It was Otto Loewi who said this in 1959, at a time when hardly anything was known about the role of cellular calcium, apart from the general idea that it had some important function in muscle contraction. The phrase obviously shows the very remarkable clairvoyance of Loewi: Now, four decades later, calcium has pervaded all areas of cell regulation, and is recognized as the most versatile of all cellular messengers. The discovery by S. Ebashi of Troponin C as the first intracellular calcium-binding protein which is essential for the contraction of muscle was the landmark contribution that started the ball rolling. Next came the EF-hand concept by R.H. Kretsinger, which gradually made it clear that calcium mediates its messenger function by binding to a special class of proteins, the EF-hand proteins: close to 600 of them are now known.

This volume discusses a list of topics which illustrate the regulatory function of calcium on a variety of extra- and intracellular processes, concentrating on the role of binding proteins to it. The introductory chapter by R.J.P. Williams develops the concept of calcium homeostasis and discusses its evolutionary aspects based on a controlled system of dynamic balanced flows. R.H. Kretsinger, S. Nakayama and H. Kawasaki then discuss the evolution of the EF-hand proteins, which they propose developed from a common precursor. The structural aspects of the most important intracellular calcium-receptor protein, calmodulin, and its interaction with targets, are discussed by T. Yuan, K.L. Yap and M. Ikura. Next, the relatively new topic of extracellular proteins containing a calcium-binding EGF-like domain is described in the chapter by A.K. Downing, P.A. Handford and I.D. Campbell which stresses the influence of mutations on a number of inherited diseases.

The second portion of the book deals with the role of calcium binding proteins in a variety of calcium-dependent signal-transduction pathways, including gene expression, which has recently been shown to be regulated by calcium/calmodulin-dependent kinases and phosphatase(s). This is illustrated by two chapters that cover the role of the kinases (J. Krebs) and of the calmodulin-dependent phosphatase calcineurin (C.B. Klee). The final three chapters of the book deal with protein systems controlling the intracellular level of calcium. G. Inesi and C. Toyoshima discuss the structure-function relationships of the calcium pump of the sarco- and endoplasmic reticulum. The plasma membrane counterpart of this ATPase is the calmodulin dependent calcium pump, which is discussed by D. Guerini. The two pumps are responsible for the fine tuning of intracellular calcium since they interact with it with high affinity. However, most cells also contain high capacity, lower affinity calcium-transporting systems. The most important of them is the sodium-calcium exchanger, which is the topic of the final chapter of the book by N. Gabellini, A. Zatti, and E. Carafoli.

Padova Zurich,
February 2000

Ernesto Carafoli
Joachim Krebs

"Yes, calcium, that is everything ..."

Contents

Calcium Homeostasis and Its Evolution

Robert J.P. Williams

Inorganic Chemistry Laboratory, University of Oxford, South Parks Road, Oxford, OX1 3QR, UK
E-mail: Susie.Compton@chem.ox.ac.uk

The homeostasis of calcium is managed very differently in different organisms. A general outline of homeostasis, a dynamic balanced flow, is therefore described before a description of the systems in particular classes of organism. The organisms are treated in the sequence: prokaryotes, single-cell eukaryotes and, lastly, multicellular organisms reaching up to man. The importance of local homeostasis in compartments contributing to the whole flow balance and of the environment are stressed. Homeostasis can only be managed by the continuous use of energy and, for any one element, here calcium, has to depend on the balanced flow of many others.

Keywords. Homeostasis, Cell compartments, Calcium gradients, Calcium flow, Coupling to calcium, Mitochondria, Endoplasmic reticulum

1 Introduction

Homeostasis is a condition of a cell which describes a steady state of flow of material and energy. It is therefore a dynamic concept always relating to some energised condition, so that in different steady states a cell may be in periods of quite different flows of material and energy. For example, bacteria may be in either a resting (lag) or a growth (log) phase. In both cases there is an underlying steady metabolism but they are very different in that there is very little flow in the lag phase. Conditions in cells of higher organisms can also be at rest, in growth or in excited active states. This is seen on the large scale in dormancy as opposed to in the fully aware states of some animals, and in the difference between sleeping and waking states of the brain. In advanced organisms it is not only the state of a cell or the inputs to it and outputs from it which are important but control over different cellular environments and over extracellular fluids also need to be exerted. Both the cells and the external fluids are in separate homeostatic conditions. The essential feature which we must observe is that, with reference to any substance, here especially the metal ion calcium, its concentration is held constant in a given steady state for some period of time due to its fixed input and output, no matter in which compartment it is, with constant use of energy. Thus the concept of homeostasis is quite different from thermodynamic buffering which does not include flow of material or a constant use of energy.

To reduce the complications of many compartments we shall often describe the homeostasis of calcium by reference to fixed levels of free Ca^{2+} ions in a given compartment and assume for the most part that this level is in local compartmental balance or equilibrium with all possible calcium-binding sites. (In fact it is in not more than a steady state). Here a compartment includes separately the cytoplasm, for all organisms, and then vesicles and organelles, and finally the external fluids of multicellular organisms as we proceed from prokaryotes, to different eukaryotes of increasing cellular complexity. In each case there is an energy input difference between the compartments so each has its own "equilibrium" or steady state. In other situations we may need to refer to a fixed gradient of calcium ions across a compartment apart from gradients across membranes which are always present.

Calcium-binding sites are exchanging Ca^{2+} all the time using input and output channels, exchangers, pumps and carriers. This flow of calcium therefore generates a current which varies around selected regions of a cell, see below. Any current is related to some gradient of course. This somewhat unusual approach is quite different from a conventional analysis of the functional value of calcium based on studies of the in vitro activities of isolated calcium-binding proteins or other molecular units. Again there will be conditions in which fluctuations in energy, due to changes in external circumstances, including signalling and acting on controls over pumps, exchangers and channels, will cause changes in calcium flow, i.e. loss of homeostasis. We shall not analyse here short, single pulses of the cation lasting less than one tenth of a second. However, a continuous series of short pulses with short, but rather longer, recovery times will produce an elevated long-term steady-state concentration of calcium underlying the pulses and this change does relate to a changed homeostasis (Fig. 1) [1, 2]. On the contrary a short-term burst of activity

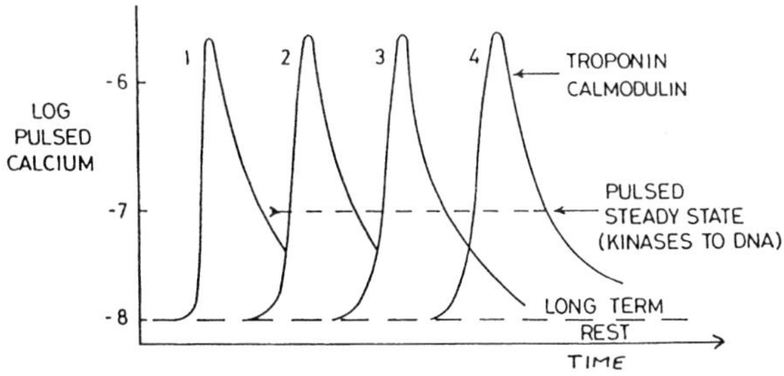

Fig. 1. Schematic diagram showing how a single fast pulse can cause changes in tension (troponin) and enzyme activity (calmodulin) but not affect kinases relaying to DNA thus leaving homeostasis unchanged, while a train of pulses also raises the calcium level to a new homeostasis by relaying information more slowly to DNA via kinases and phosphatases [1, 2]

followed by a resting period does not give a long-term elevation of calcium concentration, see below. Of course the extent of all input does depend on the external supply of calcium.

Given that homeostasis is related to many kinds of flow, and hence to energy supply, it is obvious that in a cell there has to be a relationship with the controlled use of energy transduction within the cell which therefore has to be connected to calcium gradients and to those of many elements other than calcium. We cannot expect fixed flow based on exchangers (Na^+/Ca^{2+} or H^+/Ca^{2+}) and/or on ATP-ase pumps to be other than dependent on homeostasis of Cl^-, K^+, Na^+, H^+ and ATP concentrations and they must be controlled individually, yet cooperatively, as opposed to the primary general external sources of energy, light and dioxygen, which are not controllable. We must pursue this problem and show that ATP and H^+ levels are dependent on management of (1) supplies of reduced organic compounds such as sugar (glucose), which does depend on light in some organisms (plants), and (2) on O_2 supplies (also in flow) in reaction with products from sugars in most organisms, and hence we need to show how these dependencies are related to Ca^{2+} levels and flows. We know of course that these reaction processes depend on the presence of controlled amounts of many elements. The implication is clear: there has to be feed-back and feed-forward relationships involving many elements to establish homeostasis between energy input to and reactions of any given element such as calcium.

A further pre-requisite needs to be analysed – the source of proteins with which calcium interacts. Here we must examine gene expression, see for example [3]. Proteins in cells are constantly turned over and therefore maintained synthesis of them is required. The need is for *constitutive* synthesis of many of the essential proteins which is limited by growth conditions. Such synthesis is under hormonal control in higher animals, thereby depending on the well-being of the organisms generally. Again we can only indicate ways in which homeostasis of Ca^{2+} is linked back to protein production via transcription factors and here elements such as Mg, P, Fe and Zn become critical. In such circumstances we shall again be forced to describe features of homeostasis which involve elements other than calcium while describing that which relates to calcium directly. The various factors involved in calcium homeostasis are listed in Table 1. Since the number of these factors has increased vastly during evolution this review will follow what we take to be the progress of evolution (Fig. 2 and Table 2) in order to look at the simplest Ca^{2+} functions first. We shall start therefore with the essentials of homeostasis in a primitive prokaryote cell, progress to single-cell eukaryotes with organelles, and then to multicellular organisms. Of course in this progression homeostasis becomes increasingly complicated by compartmental divisions but we consider that, of necessity, evolution develops by adding on novel features to pre-existing ones (while even losing some) (see Table 3) and of necessity networking them all together. Thus, knowing in outline the homeostasis of the simplest cell even where calcium involvement is slight will give us a firm basis for homeostasis in more and more complicated cells.

Table 1. Factors controlling calcium homeostasis

1. External availability of calcium. In so far as availability involves controlled extracellular fluids, as in many multicellular organisms, it is assumed that these fluids are homeostatically controlled and they are treated as internal to the organism. *External calcium* is then that concentration free in the environment, e.g. the sea
2. Application of energy. Here we are not concerned directly with calcium gradients but with the total energy uptake of a cell since in homeostasis calcium gradients have to be in balance with those of many other elements. However, if homeostasis is lost then extra energy is required to return to the homeostatic condition. In many organisms this restoration is dependent on calcium-dependent switches much as the loss of homeostasis involves changes in calcium levels
3. Production of molecules interacting with calcium, including reactions giving precipitates. This includes pumps and channels in membranes, carriers and binding proteins which bind calcium in compartments, as well as hormonal factors which regulate expression of them. We must understand also how calcium and other material concentrations are related to gene expression of all these proteins in particular cells through feed-back

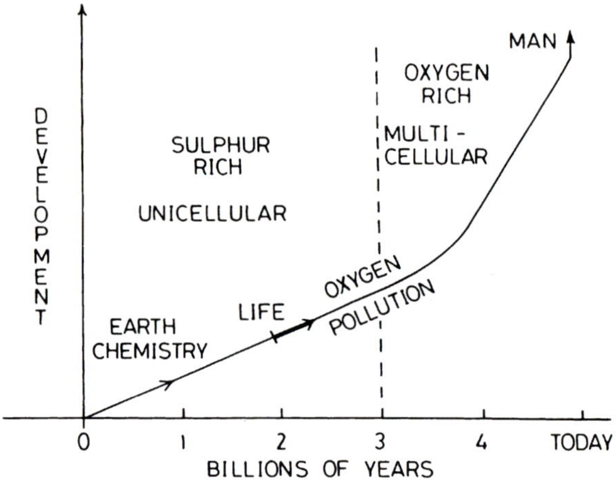

Fig. 2. An outline of evolution. Prokaryotes have always existed while single-cell eukaryotes developed some 2×10^9 years ago. Much of this development was anaerobic. Multicellular organisms arose with the dioxygen in the atmosphere giving rise to new organic chemistry [3]. The utilisation of calcium gradients was coincidental with this cellular development (see text)

Table 2. Calcium in evolution

Cellular organisation	Calcium function
A. Prokaryotes Single cells	Externally: wall crosslinks; random mineralisation Internally: very slow signal: swimming. No message to genetic code except for sporulation under starvation
B. Early eukaryotes Single cells (no organelles)	As for (A) but external events relayed to internal structures. (a) shape response: contractile devices (b) metabolic and energy response (c) increased response from endoplasmic reticullum stores (d) controlled external mineralisation (e) control over cell death
C. Late eukaryotes Single cells + organelles, e.g. see *Acetabularia*	As for (B) but (a) extra response and energy release from mitochondria and chloroplasts (b) necessary connection to genetic code through phosphate compounds to give growth, see (C), (a) (c) growth pattern linked to constant circulating calcium current around cell edges: differentiation
Early multicellular nimals (nematode). ants are similar but hout nerve structures	As for (C) except (a) link to nerve response as well as chemical response to the environment (b) control of circulating fluids inside organism by organs, releasing hormones (c) connective tissue allows buildup of internal mineralisation (d) protective and immune system organisations
ulticellular with brains	As for D except (a) the brain is a new plastic organ (unlike other organs) and calcium currents around nerve cell tips of the brain now linked to local growth to give memory. Apoptosis

les of maintained pathways throughout evolution

	Example
gradation	Glycolysis
reaction series	CO_2 incorporation, later reversed to energy capture, Krebs cycle
	Methionine initiation
ses	Nucleic acid pathways
is	β-carbon oxidation/reduction using flavin

2 Homeostasis in Primitive Prokaryotes

We do not know the nature of the earliest life; however, we do know the basic rules of the simplest living organisms of all kinds as they exist today. Certain features of today's simple organisms, which we surmise resemble those of primitive life, can therefore be stated clearly [4].

1. All life is cellular, i.e. is contained by membranes. There is an internal cytoplasmic fluid.

2. The overall metabolism is based on seven non-metal elements, H, C, N, O, P, S and Se. With these elements all the major polymers of all cells are made. Hence the major metabolic pathways involve them. Their uptake and loss are the major flows of material. Most of this metabolism in the cytoplasm is unchanged in all cells to this day (Table 3).
3. The major flows of energy are through ion gradients and pyrophosphate in tri-nucleotides, especially adenosine triphosphate (ATP).
4. In order to link energy and material flow in pathways at reasonable rates the cellular systems require catalysts. The earliest catalysts were not just organic molecules including polymers but required at least Mg, Mn, Fe and some Zn as well as perhaps some Ni, Co and Mo (W). We believe Cu and elements such as I were only used later, since they require oxidising conditions.
5. The maintenance of the stability of cells, osmotic, electrical and chemical, requires the cell to reject certain elements, all as ions, namely Na^+, Ca^{2+} and Cl^-, while retaining K^+ and Mg^{2+}. The levels of these simple ions are related to the cell's activities, both to metabolism and to the functioning of DNA. Here we note especially, and in the context of this book, Ca^{2+}. This requirement to reject Ca^{2+} in the initial stages of life is the pre-requisite of all its advanced functions in higher organisms. Later Cu is also largely forced to be extracellular.
6. We shall only deal with living reaction systems which are under genetic, DNA control, even though the most primitive systems of flowing chemicals in cells may not have been so coded.
7. To maintain steady states of flow, cells devised numerous signalling (circuit) systems employing carriers and messengers, amongst which co-enzymes are of major importance.

As stated earlier, to describe cellular homeostasis we may have to examine the chemistry of about twenty elements (Table 4). [There is some difficulty in the use of words in the above description. It must always be remembered that the use of words in sciences is for operational purposes and that a word may well change its meaning with time as new discoveries are made. Different authors refer to similar but not identical concepts by the same name. A gene in this book is a sequence within DNA but its relationship to the original definition of a gene as an inherited piece of chemical machinery required to produce a characteristic is not straightforward (see [4]). Definitions of types of genes also vary somewhat so we present no more than a way of looking at them in Table 5].

3
The Role of Calcium in Primitive Prokaryotes

The use of calcium in early prokaryotes is probably similar to its use today in bacterial cells such as *E. coli.* (Fig. 3). The *E. coli* bible [5] gives very few references to calcium. It is clear that the internal level of Ca^{2+} is close to10^{-7} M in these modern bacteria but this value is not very critical to their survival and calcium levels may well have been as high as 10^{-5} M in the most primitive cells. Higher calcium levels than 10^{-7} M in modern simple cells promote some new activity, mainly disoriented

Table 4. Involvement of elements in homeostasis during evolution

Primitive anaerobic prokaryotes	Early (anaerobic) single-cell eukaryotes			Later and aerobic multicellular eukaryotes
H, C, N, O, P, S, Se substrates and polymers		→		
H^+, Na^+, Mg^{2+}, Cl^-, K^+, Ca^{2+} exchangers		→		
	ATP-ases, Ca^{2+} signalling		→	calmodulin, annexin
H^+, P, S, Fe signals		→		
Mn, Fe, Mo, Se		→		in cells and high-potential enzymes outside cells
low-potential enzymes		→		
Ni enzymes (H_2, CO)		→		
Ni(urease)	→		-->	plants only
Co(B_{12})	→		-->	animals only
Zn(enzymes)		→		in vesicles
			→	Zn signalling
			Cu →	Cu enzymes (outside)
Outer membrane		→		
	Vesicles and inner filaments and signalling		→	Iodine hormones
			→	Outer filaments and signalling between cells Na^+/K^+ ATP-ase organic hormones

Table 5. Definitions of genes with examples in *E. coli*

Definition	Pathway example	Metal distribution example
Constitutive or constitutive with repression	Glycolytic path and early H^+/ATP synthase Anaerobic metabolism	Na^+/H^+ pump K^+ channel Ca^{2+}/H^+ exchange Cl^-/H^+ exchange
Constitutive but with (end-product) repression	Amino acid synthesis Nucleotide synthesis	Ca^{2+} channel(?) Ca^{2+} ATP-ase(?)
Induced	Organic drugs Oxidative metabolism	Cd^{2+}, Hg^{2+}, Pb^{2+} pumps (outward)

Constitutive genes may be expressed synchronously with the cell cycle or at a particular rate limited only by feed-back from a gene product. These gene products are always present.
Those genes which are inducible are blocked normally but this blocking cannot be 100% since blocking is due to finite binding of a repressor molecule which has a binding constant allowing the repressor to be off the DNA every now and then and therefore generating a very low level of expression.
In higher cells the constitutive genes of parent cells can be repressed on differentiation due to novel production of repressors. This allows cells of very different kinds to be present in one organism and allows all cells to remain in a steady state without growth.
Some genes are for regulation only and are not expressed. There are also genes for the binding of polymerases.
Clearly all the above "definitions" are but operational (not definitive!) and, depending on cell type, a gene for a given protein may be in any one of the above classes.

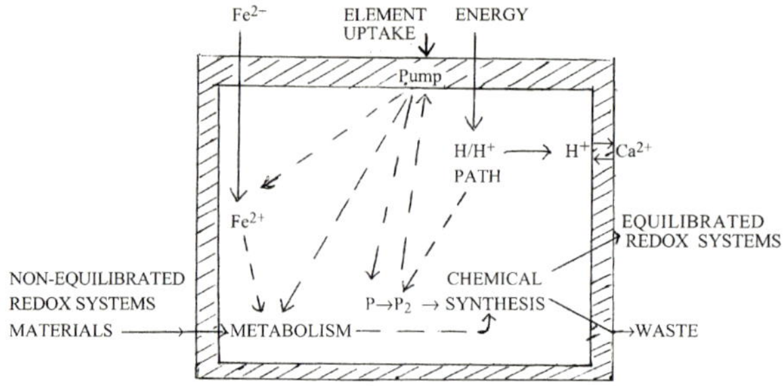

Fig. 3. A scheme of prokaryote dependence on energy and its utilisation in the uptake and rejection of many elements. Calcium is simply rejected by a proton exchanger. Note the need for Fe, H and P

swimming [6]. Their calcium proteins or genes are not well described. There certainly is a calcium/proton exchanger which acts to keep calcium internally at a low level. Generally speaking, exchangers, e.g. $2H^+/Ca^{2+}$ or $3Na^+/Ca^{2+}$, are high-capacity, low-affinity systems while Ca^{2+} ATP-ases, see below, are of low capacity and high affinity. Thus exchangers probably pre-date many ATP-ase pumps (Fig. 4). Another possibility is the use of a polyphosphate channel system for Ca^{2+} controls, as recently described [7]. There is no evidence for any of the well-known calcium proteins of yeast and higher organisms such as calmodulins, annexins or S-100. It is probable that Ca^{2+} homeostasis is a rough and ready affair and it is certainly of slow response time [8]. Calcium is important in the coat of *E. coli*, the outer membrane, and very low external calcium induced by the presence of EGTA removes the coat and seriously impairs the life of the organism. Higher calcium levels than 10^{-5} M internally kills the cells or causes them to "hibernate". Thus under stress many bacteria sporulate, allowing calcium to enter the cell and then use calcium to form a strong outer coat of a dipicolinate complex around the DNA and a limited set of other chemicals to form a spore. The influx of calcium may activate this process – a failure of homeostasis due to low energy status in the lag phase, which may be likened to apoptosis in higher cells. The binding of Ca^{2+} to dipicolinate requires a free $[Ca^{2+}]$ of some 10^{-4} M, now in the "dying" cell!

While *E. coli* may not have a Ca^{2+} ATP-ase as other prokaryotes do, and as stated it may well be a later addition in evolution, the Ca^{2+} ATP-ase pump works down to a

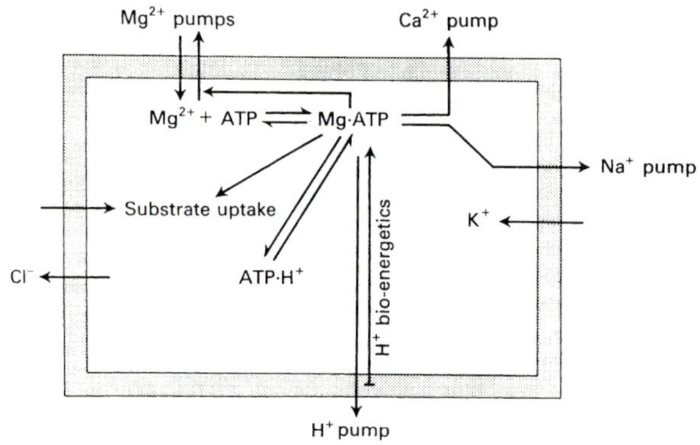

Fig. 4. The probably later addition of rejection of calcium by an ATP-ase pump and the need for Mg for its homeostasis. Note the rejection of Na and Cl with uptake of K

somewhat lower Ca^{2+} concentration. There is the presumption that the outward pumping of [Ca^{2+}] from the cytoplasm is managed more strictly in later prokaryotes. In particular, in certain photosynthetic bacteria, it may well be that calcium is already important in controlling the release of dioxygen, restricted by a necessary connection between light input and use of energy to make reducing equivalents and ATP rather than to store energy as release O_2.

It appears safe to conclude that Ca^{2+} ions, though held at a low level, perhaps initially no lower than 10^{-5} M, did not play a major part in primitive life [9]. The homeostasis of the cell was of course linked to a variety of connected levels of other elements. In the presence of adequate sources of H, C, N, O, S, P and Se to build cell chemicals, control probably rested in energy management. Here two processes dominate: (1) the generation of a proton gradient and (2) linking this gradient to ATP production in part and in part to uptake of essential chemicals and rejection of others (Fig. 4). The generation of a proton gradient required especially appropriate amounts of iron (Fe^{2+}), levels for electron transfer and the ATP production depended on controlling H^+, Mg^{2+} and phosphate in the cytoplasm. The four were readily linked through complex formation, Mg^{2+} and H^+ with phosphates, while chemical redox reactions linked Fe and H, and there were also feed-back connections of all four through metabolism of carbon compounds and genetic controls. Thus we can consider that a network of carriers of the fundamental building blocks of biological polymers and of energy, the co-enzymes, fed and controlled (through feed-back) inter-communicating pathways of metabolism so as to establish primitive homeostasis which at requisite energy input also set a low homeostatic Ca^{2+} concentration [3].

The whole required structural and catalytic units, proteins, which were supplied or restricted by feed-back to DNA and RNA coded production machineries. Each of the major pathways of the four elements, Fe, H, P, Mg, as well as of C, N, O, S and Se, were therefore linked through genetic regulation as well as through chemical controls to one another and then to all other cell activities. For example, two major regulatory routes are through the transcription factors of the *Fnr* and *Arc* type genes, which represent redox-active management related to the Fe^{2+}/Fe^{3+} balance or redox poise, through the protein Fnr, and magnesium-dependent phosphorylation related to [ATP]/[ADP, P] poise, through Arc, a kinase [5]. H^+ probably affects both types of gene. We see that the earliest cells must have relied on an internal homeostasis of at least H^+, Fe^{2+}, Mg^{2+} and phosphates apart from systems related to C/N/O/H/S/Se metabolism. We shall refer to this as $Fe/H^+/P/Mg^{2+}$ homeostasis when the input of organic material and energy are fixed and for constant growth conditions. For connections to other activities, see Table 4. Changes in material or energy supply affect the $Fe/H^+/P/Mg^{2+}$ status of the cell and expression of proteins related to it. As stated calcium played little or no part in this homeostasis being maintained low in the cytoplasm at the expense of energy by either H^+ or Na^+ exchange.

It must be appreciated that the flow of material and energy in the most primitive cells was most probably random inside the cytoplasm though, as cylindrical cells developed, the flow may have become different toward the poles and therefore had a vector quality within the cells as well as across membranes.

4
Homeostasis in Unicellular Eukaryotes

The yeast genome is now known completely and for present purposes it will be assumed that yeast is a typical unicellular eukaryote. Much though this may be a sweeping over-simplification it is probably a reasonable assumption for an understanding of the appearance of certain major calcium functions. It is important to note here that the development of the eukaryote cell with its separated nucleus came with four other major changes in cell structure (see Fig. 5):

1. Cells increased greatly in size,
2. The outer cell membrane became flexible,
3. The cell now included vesicular and organelle compartments so that activities in the cell were segregated. (Only vacuoles are found in bacteria), and
4. The shape of the cell and the disposition of compartments depended on an extensive filament structure *internally*.

To understand calcium homeostasis, and now calcium-trigger functions in these organisms, we have to keep in mind firstly that the major internal activities of cells as seen in the cytoplasm of prokaryotes, such as general metabolism and synthesis (Table 4), could not be greatly adjusted. Calcium functions, to be described, then had to add on to previous $Fe/H^+/P/Mg^{2+}$ homeostatic functions but be interwoven in a network with them.

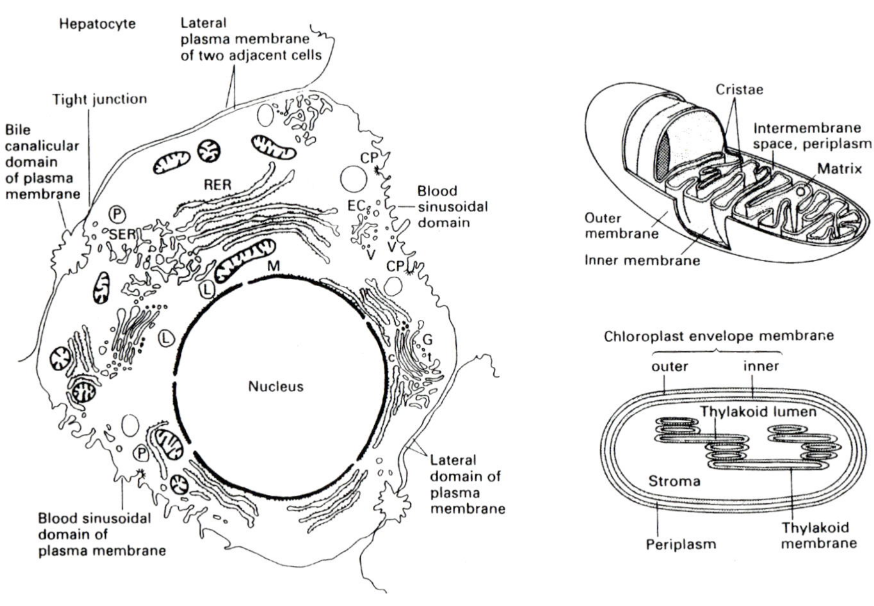

Fig. 5. Cellular structure of an advanced eukaryote (here a hepatocyte) indicating the complexity of compartments. Note the flexibility of the membranes. Filaments are not shown. The organelles in plants (chloroplasts) and in plants and animals (mitochondria) are shown on the right. Calcium acts as a major communicator between the compartments. It is current opinion [10, 11] that the mitochondria and chloroplasts are very long structures weaving through cells

The distribution of calcium in the cell is now important. Under steady-state growth conditions calcium is close to 10^{-8} M in the yeast cytoplasm and its mitochondria. The internal fluids of the organelles are cytoplasmic in that they resemble prokaryotes being almost free from Ca^{2+} at about 10^{-8} M but, unlike the eukaryote cytoplasm, organelles are not very sensitive to considerably higher levels. Various vesicular systems have higher calcium levels up to 10^{-3} M, including some vacuoles (the Golgi apparatus), but in particular the endoplasmic reticulum. These vesicles are effectively extracellular in calcium content. Homeostasis is now a linkage not just of internal cytoplasmic $Fe/H^+/P/Mg^{2+}$ systems and connections to other element activities but also to the external world, the interior of vesicles, the tension in filaments, and the internal organelle solutions. Unlike a prokaryote, a eukaryote has a very flexible response to the environment. It is the required additional connection between all the new structures and the new compartments which is managed in considerable part through Ca^{2+} ion flow. There are then different homeostatic conditions based on this flow, using novel calcium-binding proteins (Table 6) in the different compartments as well as in different cell states. Very unfortunately for our understanding, the distribution of calcium in the larger cells of eukaryotes may not be uniform in the membranes and even in the cytoplasm and is as yet little understood, though it is a major feature of homeostasis. The situation arises from the asymmetric distribution of Ca^{2+} inputs and exits around the very asymmetric cell and its interior. Before we describe this flow, however, we will analyse a steady state as if it were an equilibrium condition uniform in each compartment.

In eukaryote cells as primitive as yeast we see the appearance of some new classes of calcium-binding EF-hand proteins including calmodulins (Table 6) [12]. The calmodulins have binding constants of about 10^{-7} M^{-1} for Ca^{2+} and are connected to phosphate metabolism through the activation of both protein phosphorylation (kinases) and to phosphatases (calcineurin). They always possess some basal level activity which is altered on the rise in Ca^{2+} concentration. The phosphorylated or in some cases the dephosphorylated proteins are transcription factors so that a given calcium concentration is linked to a given activity of the cell. [Only later perhaps is calmodulin linked to cell filament tension (through a protein called troponin)]. As the single-cell eukaryotes became more sophisticated so the number and variety of these proteins increased in the cytoplasm and in membranes. Notice that because all the Ca^{2+}-binding proteins in the cytoplasm and in the membrane pumps have to respond to relatively small changes in Ca^{2+} concentration, they must all have

Table 6. Calcium proteins in yeast[a]

Protein	Number
Small EF-hand proteins	5
Calmodulins	1
Annexins	0
S-100 (Calbindin)	0
Channels	2
Pumps (ATP-ase)	2
Exchanger H^+/Ca^{2+}	1

[a] These data were kindly given by Prof. T.N. Davis, Dept. of Biochemistry, University of Washington, Seattle, USA. It does not include calcium binding associated with filaments or on external surfaces.

approximately the same binding constant. This is of the essence in homeostasis and switches to different such steady states *which must use "equilibrium" exchange.* It is differential use of energy and material input which is reflected in different Ca^{2+} levels.

One notable feature of eukaryotes is the dependence of the activity of organelles, mitochondria and chloroplasts on calcium [11]. The mitochondrial dehydrogenases (present in yeast) are calcium dependent as is the release of dioxygen from chloroplasts. Here is a direct feed-back link between energy and calcium concentration. We return now to the calcium flow between all compartments which is continuous at a low rate but can be increased by environmental events or changes in cell protein content during a cell cycle.

The eukaryote cell has active calcium input channels and exit pumps in its outer membrane (Table 6). Inevitably there are always small background calcium currents flowing around these sites in membranes (Fig. 6). A similar flow is observed between organelle and cytoplasm and between vesicles and cytoplasm and between organelle and vesicles such as the endoplasmic reticulum. Obviously, the flows are most active around the membrane areas which are adjacent and have both the most input and output channels and pumps. Since most of the membranes are of variable curvature we must not expect equal flow everywhere since the curvature will affect the distribution of channels and pumps and of all other membrane proteins. The connection of Ca^{2+} flow is therefore to all features of the cell locally since calcium is now linked to metabolism (calmodulins), to internal filament structures (myosins) and so to tension, to the shape of membranes and to the release of chemicals (exocytosis). The condition of the cell is represented by all these features. Homeostasis can be viewed as *a fixed pattern of flows* of calcium with localised activities of the cell in a what we could call a dynamic mosaic (Fig. 7).

Let us attempt to summarise the effect of calcium on homeostasis going from an *E. coli* to a yeast cell. The *E. coli* internal system symbolised by the $H^+/P/Fe/Mg^{2+}$ status is virtually independent of calcium. To a large degree it is not strongly interactive with its environment except that it can metabolise and grow at different rates. Its lifetime can be as short as 30 min and its cytoplasm is, during this time, just protected from calcium. Now a yeast cell has most of the same internal cytoplasmic features but it has an advantage in that it can sense the environment and can feed off prokaryotes and their chemicals. It has the disadvantage of a longer lifetime. Hence it is dependent not only on utilisation of the environment but also on protection from it to a degree unknown in prokaryotes. Both sensitised conditions were made possible by the use of the calcium gradient, already generated in prokaryotes, and seen today in *E. coli*, as a source of information, messages, about the internal and external environment. Thus by having calcium-responsive inlets (channels) for sensing environmental content (danger, food, binding surfaces, etc.) it could use the calcium to generate all kinds of response in the cytoplasm, e.g. to stimulate metabolism (a kinase), to cause contraction or extension locally, or to release digestive or poisonous chemicals from vesicle stores.

The calcium message could be amplified from the endoplasmic reticulum and could also be used to stimulate greater activity of the organelles to provide energy. At all times excess calcium could be rejected by the exchangers or ATP-ase pumps, either to the external environment or to vesicles. The active, so-called *resting*, state of

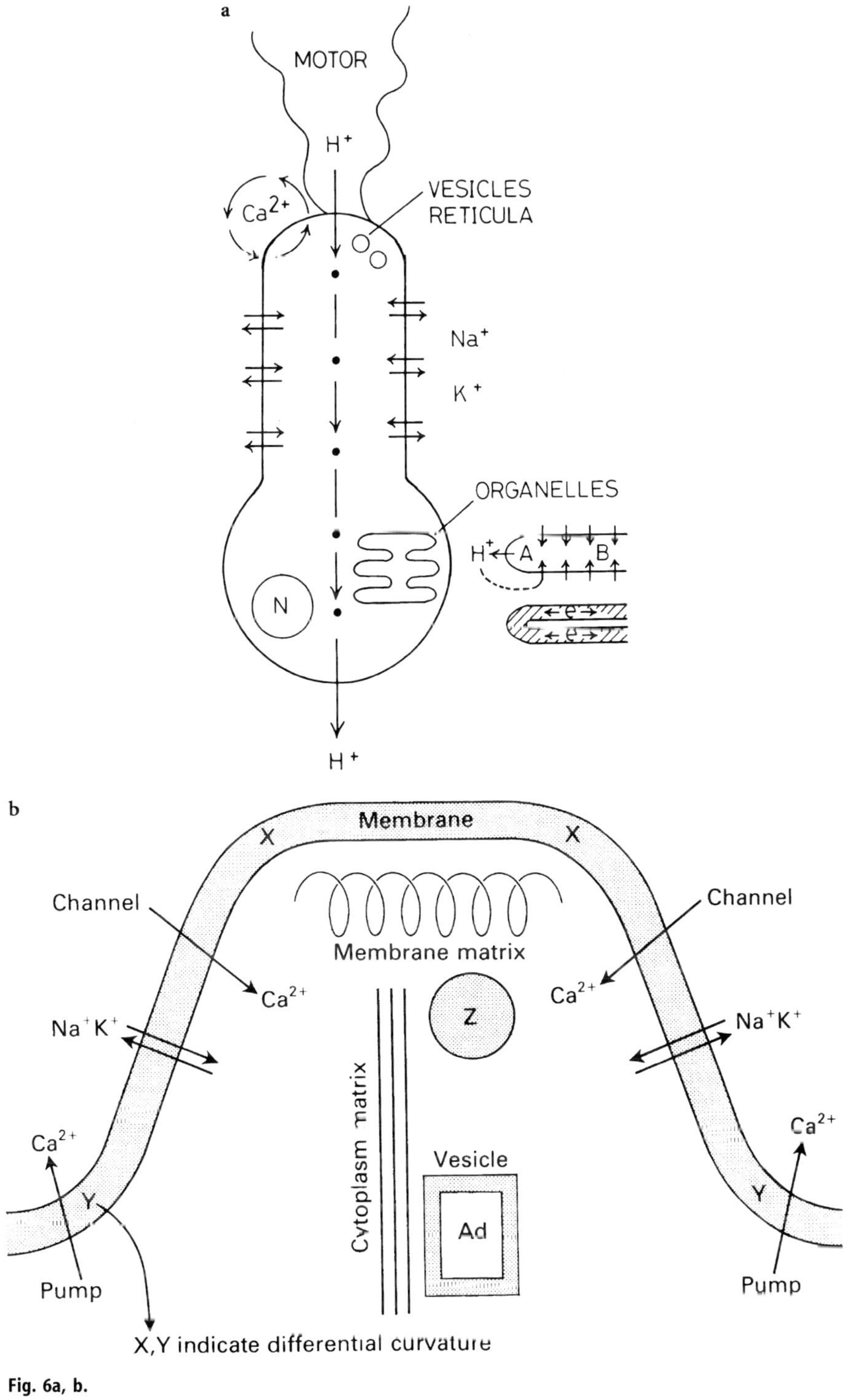
a
MOTOR
H+
Ca2+
VESICLES
RETICULA
Na+
K+
ORGANELLES
H+
A
B
e
N
H+
b
Membrane
X
X
Channel
Channel
Membrane matrix
Ca2+
Ca2+
Z
Na+K+
Na+K+
Cytoplasm matrix
Vesicle
Ad
Ca2+
Ca2+
Y
Y
Pump
Pump
X,Y indicate differential curvature

Fig. 6a, b.

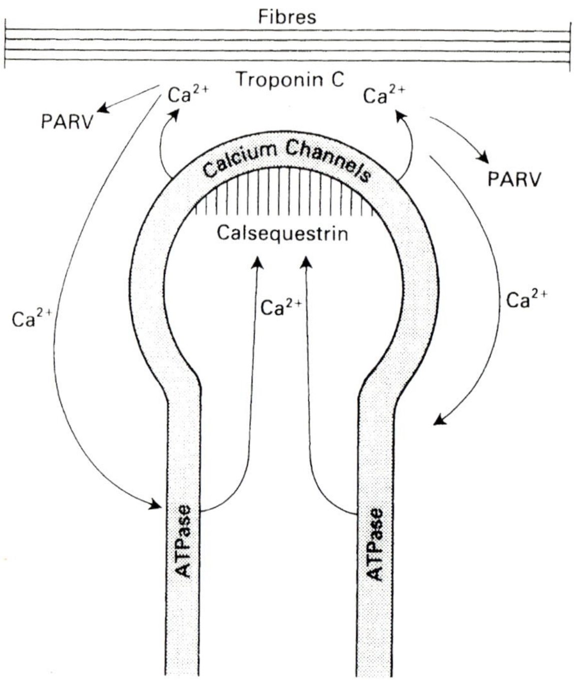

Fig. 6a–c. Flow diagrams of calcium ions across the outer membrane of cells to give a sensing current of the outside environment. **a** A simple eukaryote cell showing the Ca^{2+} flows and those of various other ions. Homeostasis now connects to the environment but does not control it. **b** A more detailed picture of the disposition of Ca^{2+} pumps and channels around a cell tip. **c** The most sophisticated Ca^{2+} flow in an animal muscle system with a controlled environment

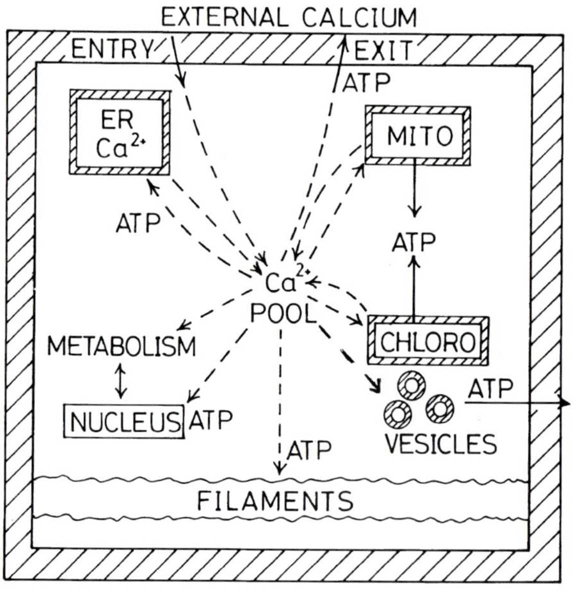

Fig. 7. Flow of calcium ions between compartments and activities in the cytoplasm as well as to the outside (see Fig. 6). Homeostasis is monitored at all levels by calcium

such a cell is always receiving minor inputs and generating minor export of calcium so that a circulating current of calcium ions flows around all the membranes. Thus a variable calcium-based homeostasis is added on to that of H^+/P/Fe/Mg^{2+} with awareness of the environment. Figure 7 is an attempt to show the large number of connections between calcium ion flows and the general network of cell homeostasis which finally lead via H^+/P/Fe/Mg systems to a maintained gene expression.

Now let us impose a fixed change in the metabolic rate on the cell which could be due to exposure to a new source of materials or energy. We know that cytoplasmic homeostasis, opposite the four elements Fe, H^+, P, Mg^{2+}, changes and with it the internal chemistry with a simultaneous expression of genes, as in prokaryotes. It is essential that these *chemical changes* are relayed to the novel features of the eukaryote, i.e. to the filaments, the shape of the cell and to the organelles and to the energy generation. This changed homeostatic state again required additional signalling to the primitive Fe/H^+/P/Mg^{2+} communication network of the internal cytoplasm and this was supplied, we believe, by allowing the calcium gradient flow into the cell to change. It appears to be the case that just as change of calcium concentration assists in activated swimming behaviour to find a better environment or sporulation protection in bacteria, all be it over a wide range of [Ca^{2+}], now a small change internally, in the range 10^{-8}–10^{-6} M in the steady state [Ca^{2+}] of the cytoplasm, causes a series of changes inside eukaryote cells which go on to affect organelles and vesicles. These were not very fast responses but long-term *differentiation* changes based on environmental alterations and of necessity had to be harmonised in all cell departments. Taking the cytoplasm first in yeast we observe not just changes in the rate of metabolism due to increased calcium input but also changes in gene expression associated with kinases and phosphatases, e.g. kinase C and calcineurin. These new proteins, involved with Ca^{2+} and absent from prokaryotes, are all cooperative so that the response to [Ca^{2+}] has a high Hill coefficient. More slowly the whole cell adapts, including the flexible outer membrane, so that it loses its previous shape due to changes in filament tension. All these changes require energy. The energy demand may well be satisfied by a change in activity of the mitochondria as, in turn, their calcium level adjusts and stimulates the dehydrogenases. Chloroplasts also respond to higher calcium levels. Another form of amplification of the calcium which enters the cell via the outer membrane is through calcium-stimulated release of calcium from the endoplasmic reticulum. Thus changes in the chemical or energy supply in the outside environment are relayed to all parts of the cell. Note again that calcium here is not a rapid millisecond trigger of action in a cell but assists in a profound long-term change, differentiation.

Another novel feature we observe in unicellular eukaryotes is one of a change in the *physical responses* of the organisms. For example, an organism, such as an amoeba, may meet a physical barrier and then adapt its membrane to bind to the surface using a pseudo-pod or foot. This change requires a long-term alteration in the surface of the cell and the *local* filamentous structures related to this surface and it must be activated relatively quickly. Here the local calcium flows can be used to great advantage. Energy may be generated locally by that portion of the mitochondria closest to the contact point through a change in its calcium levels, as described above. A cell is then a mosaic of local fixed flows.

In summary, a unicellular eukaryotic cell may have a range of homeostatic states related to energy and material flow according to circumstance. The cell can revert to an initial condition and we expect all such changes to be reflected in steady states of flow of calcium, differential gene expression, and so on. The response is *a systems response* not a molecular response though it is based on molecules [3], and it is not fast.

4.1 A Differentiating Unicellular Eukaryote: *Acetabularia*

This section owes much to the work of Goodwin [13] and Harrison [14]. The organism *Acetabularia* has a simple life history. It is initially a small cell of no particular shape but as it grows it develops a protuberance, then a stem and finally a frond appears (Fig. 8). It lives in this state for some while before reproducing. The interest here lies in the dependence of the life cycle on calcium, which we can view as a continuous change in homeostatic states. *Acetabularia* growth is dependent on an external calcium concentration of around 1–10 mM in the water (sea) in which it lives. Detailed investigation shows that a calcium ion flow is maintained around the tip. Thus its development is connected to a current of calcium ions, involving both the interior and the exterior of the cell. As the tip changes shape the input and exit of calcium changes so there is a change in internal calcium continuously and of course this flow moves away from the nucleus. The changing flow then generates plasticity in the cellular system which may well depend on the link between the internal

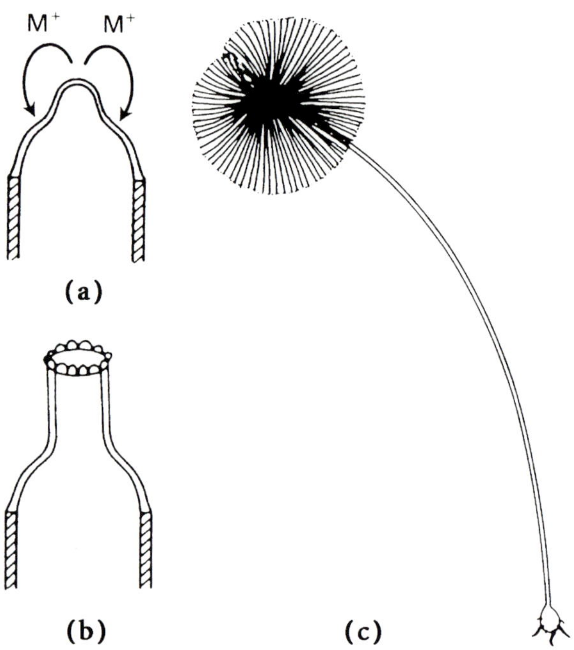

Fig. 8. An example of gradually changing homeostasis with developing structures in *Acetobularia* [13, 14]

(cytoplasmic) H^+/P/Fe/Mg homeostasis and the internal/external Ca^{2+} concentration. We cannot dissociate "homeostasis" here from development but we have to understand that development back induces a change in homeostasis so that the growth of the stalk of *Acetabularia* switches at a certain length to growth of a frond. Notice that the responses to Ca^{2+} affect protein production which have to be relayed through the H^+/P/Fe/Mg network to the genes (Fig. 9). In effect the homeostasis has become an H^+/P/Fe/Mg/Ca development pattern. Note that the description implies that Ca^{2+} is an *internal* morphogen [13, 14].

5 Homeostasis in Multicellular Organisms: Plants

The novel feature of multicellular organisms is that different cells within the organism as a whole are in different steady states or homeostatic conditions. If we look first at plant cells then we know that any cell in the organism, no matter if it is from leaf, root, shoot or flower, can become a stem cell and can produce a whole organism. The activities of genes are fixed only so long as the cell experiences a fixed environment. Quite possibly it is easier to appreciate the role of calcium in this rather flexible cellular world of plants than in animals where a differentiated cell can only revert to a stem cell under special circumstances. Now in plants calcium changes are usually not associated with very rapid response [15, 16], hence individual cells can be thought to resemble combinations of unicellular eukaryotic organisms extracted from different environmental conditions and organised in a structure related to these environments. The environment is now a product of the organism's self-generated external world within an external structure and with a new extracellular chemical flow to and from neighbouring cells and the physical contacts between cells.

Particularly novel are connective tissue and organic messengers, hormones, flowing in the extracellular fluids, which are contained by an outer structure

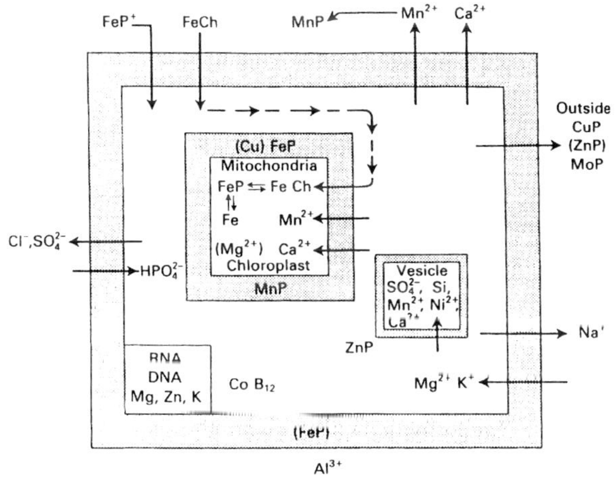

Fig. 9. Homeostasis of a eukaryote cell relates to the distribution of some 20 chemical elements including compounds of C, H, N, O, P, S, Se, as well as the metal elements shown [3]

enveloping all the cells. Initially, as cells multiply from a single cell (seed) to a group of say eight cells, the environmental factors of note are the differential polarity relative to gravity, aerobicity and light. The cells differentiate in these fields and then become affected by the neighbouring cells which form new environmental physical conditions for each another and chemical (hormone or morphogen) connections to each another. All the cells pump out [Ca^{2+}] down to around 10^{-8} M but, as the cells differentiate, there is no strict equality relative to Ca^{2+} ions between them due to differential metabolism, energy input, and so on. As the cells differentiate further towards organs, hormones are released from some cells as a steady flow which activate, through steady calcium currents, certain other cells using a fairly well-kept stationary calcium concentration gradient in the extracellular fluids, which depends on the available energy and water supply. These effects can only operate if they are linked to the pre-existing differentiated internal cellular systems, as seen in unicellular eukaryotes (Fe, H^+, P, Mg and Ca). Evolution is adding-on.

Calcium remains very much the link between the novel environmental factors, local physical effects (light), and hormone responses, to the cytoplasm and internal vesicle, organelle, and filament compartments. In particular, as hormone levels change in liquids flowing in the plant between the cells and interacting with the outer membrane of each cell, it follows that changed calcium flow into the cell will change and affect many internal properties (see Fig. 10). The control system has changed from that of the unicellular eukaryotic to an extensive linkage between all parts of the plant. The plant becomes a graded series of cells from root tip to flower in a system where the parameters of the system are like the electron in a computer circuit but are now the flows of various materials including calcium and energy. This is reminiscent of the changes in the growth pattern of *Acetabularia*. The homeostasis constantly changes with growth and perhaps with the changing internal environment of the circulating fluids of the plant. The grading in a plant is

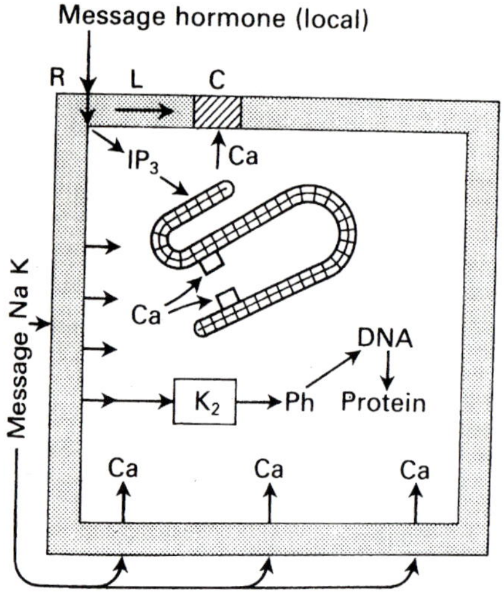

Fig. 10. The distinct ways in which overall depolarisation of a cell by a Na^+/K^+ current releases Ca^{2+} generally into a cell or hormones bind locally to a cell membrane and release localised calcium pulses, mediated by inositol phosphates (IP_3) and amplified by release of more calcium by the endoplasmic reticulum [15]. *R* Ca^{2+} input channel; *L* lipid movement; *C* outward pump; *K* kinases

easily visualised since the root is in the dark, in contact with a calcium supply, while the leaf is in the light.

In summary, we consider that major homeostatic conditions in each particular differentiated cell are as in simpler eukaryotes and are maintained selectively by a communication network of $H^+(10^{-7}$ M), $Fe^{2+}(10^{-7}$ M), phosphate (10^{-3} M) and $Mg^{2+}(10^{-3}$ M) and by various organic molecules and energy, but that in eukaryote multicellular organisms these internal messages connect to a wide range of *controlled and selective* external conditions, light, wind, hormones and substrates (due to structure), using as a connecting link the passage of the Ca^{2+} ion across membranes. We have noted how calcium became the means by which single eukaryotic cells recognised what was around them, and then utilised it to generate release of various chemicals, poisons or digestive juices from vesicles while adjusting physical structure including both endo- and exo-cytosis and growth. In multicellular organisms these vesicle-release mechanisms were converted to internal digestion and internal messages [hormones, morphogens, and transmitters (X)] which back-interacted to generate calcium signals in a distant cell. The distant cell responds but, in addition, sends a message back (F), so that the organism works as a cooperative unit.

External $\rightarrow$ Ca^{2+} $\rightarrow$ Cell 1 $\rightarrow$ release of X $\rightarrow$ X. Cell 2 $\rightarrow$ Response to Y
Activity (Y) $\uparrow$ $\leftarrow$ F $\leftarrow$ F $\leftarrow$ $\downarrow$

All such processes in plants are still relatively slow, longer than minutes rather than milliseconds.

6
Homeostasis in Multicellular Organisms: Animals

An animal has a need for fast response to environmental threats as well as to those of differential cellular development [17]. Hence the homeostatic and signalling devices of its cells require new fast modes of reaction and communication in addition to the activities seen in plants. One such mode is the development of nerves in which Na^+/K^+ currents flow, connected to muscles. These flows are not just stimulated by a threat but have a small constant background activity which is their resting homeostatic condition. Most commonly, this activity of the nervous system is discussed in relationship to the brain but, in fact, the nervous system keeps all the muscles of the body in tune. Now the nervous system has repeated junctions in its trunking. At these synapses the nerve impulse is reinforced by the release of transmitter chemicals such as acetyl choline and γ-aminobutyric acid (GABA) to stimulate the receptors of the adjacent nerves. This release is due to calcium ions which are also in constant flow around the synapse terminals, but such flows are accentuated when the Na^+/K^+ flow increases. Thus calcium ions constantly link the Na/K^+ current to the transmitter current (Fig. 10). The releases of transmitters are in packages from vesicles. The resting homeostatic condition is of a slow background flow through the whole system of nerves and contains many localised homeostatic calcium flows. On excitation, all flows are stimulated in either a short burst or to a maintained state of alertness. Hence the homeostasis of individual cells due to different flow can be of a variety of different conditions using energy differently. We must consider how this is possible in the special case of calcium.

The nervous system contacts muscles of two types: posture muscles, which maintain shape and are slow acting, and twitch muscles, which respond in a millisecond. An intermediate case is the heart [18] with a cardiac muscle which has a one second beat in man but can vary from much faster to slower in smaller or larger animals, respectively. All these muscles can change tension, the energy applied to filaments, and all these tensions are due to calcium current flow, from both the extracellular fluids and the endoplasmic reticulum (ER), stimulated by nerves, and of course to the presence of suitable calcium-binding proteins. The calcium flow can also be varied by hormonal effects, e.g. by adrenal flow from glands (Fig. 10). Thus the homeostasis of muscle, just like that of brain, is a variable of requirements. Once again the stimulus is in pulses of calcium release but *continual* rapid low-level pulses, as in slow muscles or as stimulated by a hormone, affects the calcium basal level so that while it fluctuates between 10^{-6} (maximum) and 10^{-7} M (minimum) in each burst state it never returns to the basal level of 10^{-8} M in the resting state (see Fig. 1). A twitch muscle on the other hand must respond rapidly to a surge in calcium input but then relax rapidly to its initial condition.

Clearly there is a need for the response proteins in the cytoplasm to be of at least two kinds: (a) those that respond and cause relaxation in 10^{-2}–10^{-3} s, and (b) those that sustain an activity in the background. Thus homeostasis became increasingly complicated and differentiated as multicellular organisms developed toward the vertebrates (Fig. 11). Note that the innervation of a muscle and the type of messages it receives decide the gene activity of the cell.

Nerve-nerve junctions in the brain are different from similar junctions in the main part of the body since they develop a memory of events. The memory is strengthened by repetition. This suggests that the impact of an experience on these nerve cells is to generate a permanent record. One clear possibility is that the electrical Na^+/K^+ impulses stimulate local growth of connections. Such growth could be due to the changed calcium flow at synapses. This idea was suggested some time ago [19] but only recently has evidence been gathered in its favour [20]. It may well be that nerve terminals have a special set of calcium-binding proteins which retain calcium for a longer period than fast-response proteins such as calmodulin. Growth requires a sustain message. In any event the very plasticity of the brain, unlike any other organ, leads one to postulate that it has very special Ca^{2+} proteins in its nerve cells. Note that plasticity in development associated with Ca^{2+} currents has already been observed in *Acetabularia*, see above.

Cells with rapid responses also require a rapid input of energy and the faster the responses, e.g. of a muscle cell, the more energy required. These muscles (or nerve synapses) also have to relax quickly to generate the next pulse and this too requires a supply of energy. The calcium input relaxes back to the base line to await the next need for a fast response. Energy ultimately comes from metabolism of one kind or another via ATP so that while the mechanical actions of transmitter release and actinomyosin contraction need fast action and relaxation the chemical transformations used in producing ATP can only be acheived more slowly, except for the use of sugar stores, i.e. glycolysis. We now turn to the way in which proteins

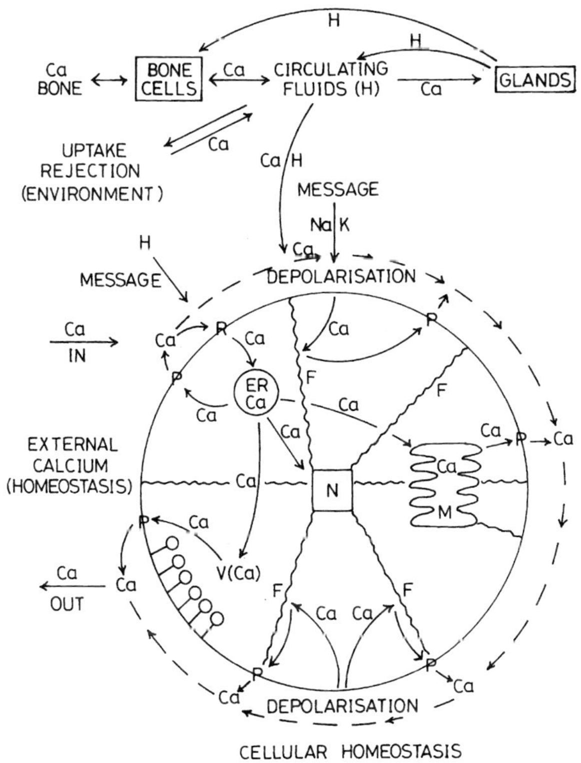

Fig. 11. Attempt to illustrate the network of calcium connections in the homeostasis of calcium in an advanced vertebrate. *N* Nucleus; *M* mitochondria; *ER* endoplasmic reticulum; *V* vacuole or vesicle; *F* filament; *P* pump; *R* rhyanodine channel; *H* hormone. The calcium concentrations, different in different compartments, are very precisely maintained

manage these homeostatic circuits involving calcium flow in energy-generating organelles.

6.1 Calcium Networking Between Compartments

The flow of calcium into the cytoplasm is coupled to the flow from the endoplasmic reticulum (ER) [10, 11]. This flow raises the level of free calcium which during relaxation is pumped out of the cell and back into the reticulum or it is pumped into mitochondria or chloroplasts (Fig. 11) [21]. The effect of calcium in these organelles is to (1) increase the metabolism of substrates into the mitochondria via activation of dehydrogenases or (2) to generate an increase in photochemical activity in chloroplasts [11]. Both these changes lead to an increase in proton gradient activity and a subsequent increase in ATP production. Due to the restricted diffusion of both Ca^{2+} and H^{+} this upgrading of ATP occurs only locally but of course it aids the rapid recovery of the cell, since ATP is required to pump calcium into vesicles (ER) and out of the cell. Other branches of the ER may be responsible for the delivery of proteins or lipids to the mitochondria or chloroplasts and again it is observed that there are local Ca^{2+}-stimulated zones for these uptakes. It is very important to recognise the dynamic or flow character of homeostasis in advanced animals.

6.2
Novel Proteins in Animal Cells

Apart from the nervous system of communication an animal has a parallel network of hormonal messages for development to that found in plants. Many of these hormones are linked to differentiation (development) and growth as they are in plants, e.g. sterols, retinoic acid and thyroxine. Very clearly with development the demand for different chemical elements changes, and this applies to calcium, and new extensive systems of calcium proteins are required even in development (Table 7) [22]. The embryo has a different demand for calcium than the growing child in such obvious features as that of bone. In the case of mammals food is supplied from the mother before the flow of material and energy is obtained by foraging. The calcium homeostasis changes during development (see Table 2).

Different types of muscle clearly require different types of protein to manage their activities. It is found that these proteins are produced in response to the level of free calcium in the cell which, as stated, is somewhat different for a muscle cell undergoing continuous low-level excitation as opposed to intermittent higher excitation with resting periods (see Fig. 1) which does not affect proteins synthesis. The Ca^{2+} free ion levels may be up to 300 nM for a cell which is continuously excited but only up to 30 nM for the second type of fast-response muscle. The connection to gene expression is made through calcium-sensitive kinases and phosphatases [23].

There is now a considerable variety of novel cytoplasmic calcium-binding proteins in the cytoplasm of advanced animal cells (Tables 8 and 9) and all of them have a part to play in homeostasis and development. One of the very striking features of evolution is the step-by-step increase in the variety and number of calcium-binding proteins needed to maintain homeostasis in differentiated states and for development (Tables 7 to 9), but, in the most advanced systems, homeostasis has to extend to the extracellular fluids.

6.3
Extracellular Calcium in Animals

There is a demand for a similar increase in the variety of and control of the extracellular proteins of circulating fluids in animals (Table 10), since Ca^{2+} flow input to cells is dependent on homeostasis of the external fluids [24]. It is a second feature of

Table 7. Some calmodulin-linked proteins

Proteins	Example of function
Protein kinases	Kinase IV; gene expression
Protein phosphatases	Calcineurin; gene transcription, amino suppressants
Nucleotide cyclases	Cyclic AMP activation
Phosphodiesterases	Deactivation of cyclic AMP
Calcium pumps	Plasma membrane transport

Van Eldik LJ, Watterson DM (1998) Calmodulin and signal transduction. Academic Press, New York.

Table 8. Types of calcium proteins in cells of higher animals[a]

Protein class	Property
Calmodulin	Fast response, many functions
(Troponin)	Fast muscle response
Annexins[b]	Moderately fast response, requires phospholipid
S-100[b]	Fast on, moderately fast off (links to filaments/membranes?)
Calbindin	Special relaxation factor
S-2 domains	Membrane associated
Calpain	Connected to apoptosis
C-2 domains	Kinase C and phospholipase activation

[a] Proteins in vesicles or organelles are not included.
[b] There are many sub-classes of these proteins.

Table 9. Metalloprotein sets in two organisms

Protein set	Yeast	Worm (*C. elegans*)	Metal
EGF, Ca^{2+}-binding Cys-rich repeats	0	135	Ca
Calmodulin-like	4	36	Ca
Nuclear hormone receptor	0	270	Zn
Binuclear GAL cluster	54	0	Zn
Metalloproteases	0	94	Zn
Mg^{2+} adhesion	4	43	Mg
Na^{+} channels	0	28	Na
K^{+} channels (voltage gated)	1	68	K

N.B. It is unlikely that any of these proteins were present in primitive prokaryotes.
Data taken from Chervitz SA et al. (1998) Science 282:2022.

evolution that control of extracellular calcium in these fluids is more and more exacting. Now the overall homeostasis in an animal can only be managed if the uptake of calcium from food is controlled. Here homeostasis is managed by a feedback relationship between calcium levels and hormones put into extracellular fluids from glands [25].

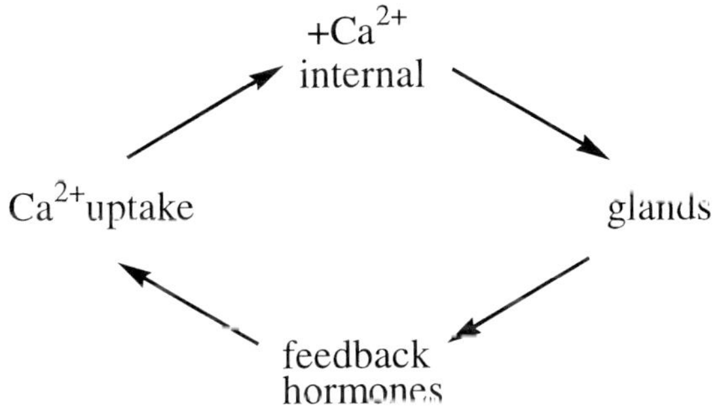

Some of the hormonal factors are well known. Calcium flow is linked to vitamin D, a sterol, which circulates to the epithelial cells and stimulates calcium flow into

Table 10. Types of extracellular calcium proteins[a]

Protein class	Property
EGF domain	Growth factors
Kringle	Clotting factors
Phosphoproteins	Carriers, mineralisation
Calsequestrin	Buffer Protein in ER
Ca-dependent enzymes	Nucleases, saccarases phosphatases, proteases
Calnexin and calreticulum	Folding of glycoproteins

[a] Proteins in vesicles are included.

the body. The functional demand of the proteins here is to transport calcium into the cytoplasm of the epithelial cell. This activity is constant to replenish calcium losses. Control is helped by the presence of the mineral bone, the solubility product of which matches the concentration product of $[Ca^{2+}][PO_4^{3-}][OH^-]$ [26, 27]. This means that in order to maintain free $[Ca^{2+}]$ the bone surface must constantly exchange ions with the solution. Management of this exchange depends on the activities of two sets of cells, osteoclasts and osteoblasts (Fig. 12). These cells have the ability to release calcium-binding proteins, osteocalcin, and special enzymes as well as an ability to adjust pH locally. Their activity must be controlled and the expression of some of the proteins is under the regulation of vitamin D. Vitamin D also acts as a hormonal messenger for the expression of the calcium-carrier protein calbindin in epithelial cells. The synthesis of vitamin D requires iron and its receptor is a zinc-finger protein. Here we see the linkage of the calcium flow to other inorganic ions, e.g. Zn^{2+} and Fe^{2+}, in addition to the three we uncovered in prokaryotes, namely $Mg/H^+/P$ (Fig. 13). Now we see that bone is not just a structure but an active functional component in the homeostasis of extracellular fluids. In all such activities there is constant feed-back, so that calcium levels are not too high or too low. The levels of circulating hormones are related to the levels of circulating calcium and vice versa.

We turn next to digestion where again we see the need for variable rates of activity in extracellular fluids and uptake. The cells here generate digestive proteins from organs such as the pancreas. The involvement of calcium ions is now in (1) activating the digestive enzymes using the homeostatically controlled calcium of the extracellular fluids, and (2) stimulating the pancreas to release the digestive enzymes

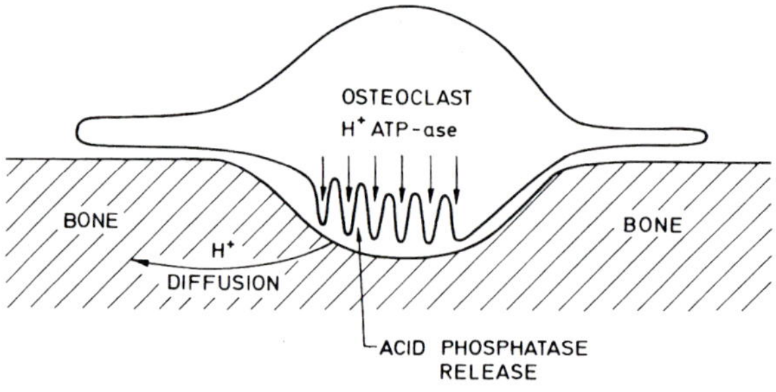

Fig. 12. An osteoclast is a special cell which releases H^+ and phosphatases so as to dissolve the surface of bone and affect the calcium levels in body fluids

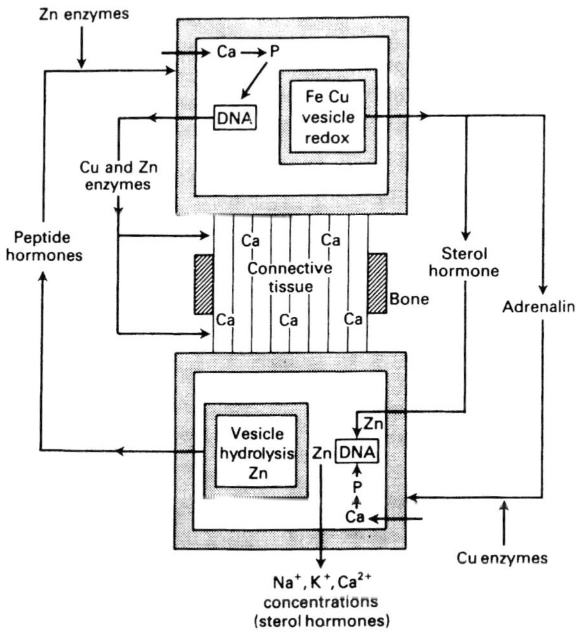

Fig. 13. The complexity of homeostasis in higher animal cells depends on elements in a different manner to that shown in Fig. 9 since extracellular levels of metals such as Zn and Cu, as well as of Ca, are involved in messages and connective tissue surrounded by extracellular fluids

from their vesicles. In the second case the input of calcium has to be changed by a hormonal signal to the cells. The cells have at least two different homeostatic conditions of calcium flow. We must turn now to the chemical mechanisms underlying the different calcium states. [A curiosity is that one external fluid, the cerebrospinal fluid, which bathes the central nervous system, has a quite different free Ca^{2+} level. Why?].

6.4 Assisted Calcium Diffusion

Ca^{2+} ions diffuse rapidly in water but of course the rate of diffusion is proportional to the ion concentration. At 10^{-8} M, the distribution of calcium by diffusion is slow. It can be helped by the binding of ions to proteins in high concentration [28, 29]. In animal cell cytoplasm there are two classes of protein which can assist the establishment of an even Ca^{2+} concentration in a cell: calbindins and parvalbumins. Parvalbumins appear to assist relaxation after a single pulse in that they can firstly rapidly buffer Ca^{2+} and then carry it to the ATP-ase for rejection. The calbindins act more as transport proteins to distance parts of a cell after local up-grading of concentration in one region. Hence calbindins are found in the very asymmetric endothelial cells in which Ca^{2+} input channels are close to the basal-lateral surface and output is at the villi. Note how *all* calcium proteins must have the same or very similar binding constants inside the cytoplasm ($K = 10^7\ M^{-1}$) so as to ensure homeostasis. This applies to calmodulins, annexins, S-100, pumps, calbindins and parvalbumins. However, rates of response may be different [18].

7 Cell Death: Apoptosis

A final feature of cell activity in higher organisms is programmed cell death [30]. All cells appear to have limits to their life-times. The general impression is that cellular metabolism slows with age and there is a corresponding slow increase in calcium concentration. At a certain stage this calcium concentration triggers internal proteases, called calpains, and cell death follows. There is a parallel here to the role of calcium in sporulation. Hence calcium is involved in homeostasis, in action and in cell destruction.

8 Summary

This chapter describes the chemical and biological value of the calcium ion. In chemistry, in vitro, our main interest is in equilibria within static, non-flowing systems. Hence we examined the way calcium forms precipitates and complex ions in solution with organic molecules produced by cells. We observed also in vitro its uses in a vast number of materials such as minerals, e.g. shells and bones. In complex formation we note that many cellular solutions contain anions which interact also in vitro with calcium ions, for example, calmodulin. The rates of exchange of calcium from bound states are known but they have little obvious implication except in vivo. Calcium ions do not act directly as catalysts of organic reactions, although they often activate the catalysts themselves.

In biological systems interest is in the above chemistry but extends to the fact that Ca^{2+} ions can carry information by flowing from one compartmental solution to another through membranes. Hence we become interested in the details of rates of calcium diffusion and exchange. The (usual) fast diffusion exchange of this divalent ion has allowed it to develop as the dominant fast second messenger. Now the flow can be looked upon as it reaches particular proteins already studied in vitro and which it activates (as seen in physical mechanical changes or chemical changes). This piece-by-piece study of cells is common. Here, however, we have chosen to stress the in vivo whole circuit of Ca^{2+} action indicating that the cell is organised both at a resting and an activated state level by the steady-state concentration of the flow of this ion. Different time constants of exchange utilising very similar types and strengths of binding internally (binding constant 10^7 M^{-1}) or externally (binding constant 10^{-3} M^{-1}), but to proteins or complexes between proteins and lipids which have different relaxation times for reorganisation.

The whole variety of systems which involve calcium have developed during evolution so that today we find only very few calcium-binding proteins in bacteria, a few in simple single-cell eukaryotes, many more in early multicellular organisms, and then literally hundreds in advanced animals (Table 8). The development is both in the dynamic homeostasis of the resting state, which is a highly active condition, and in excited states. Calcium is required to maintain the whole in a sympathetic balance and is also concerned with developmental processes. To a

Table 11. Evolution of simple ionic equilibrium signals

Stage 1	Primitive organisms Prokaryotes	$Mg^{2+}/ATP^{4-}/HP^{2-}$ controls phosphorylations; Fe^{2+} controls redox equilibrium $Na^{+}/K^{+}/Cl^{-}$ control osmotic pressure
Stage 2	Single-cell eukaryotes	Ca^{2+} controls activated states and relationship to environment; Mn^{2+} controls development of plantrelated organisms
Stage 3	Multicellular organisms	Zn^{2+} controls hormonal responses relating to growth and development and connective tissue; $Cu^{+}(Cu^{2+})$ controls connective tissue responses. Extended use of Ca^{2+} in excited states. Generation of $Na^{+}(K^{+})$ signalling and evolution of the nervous system

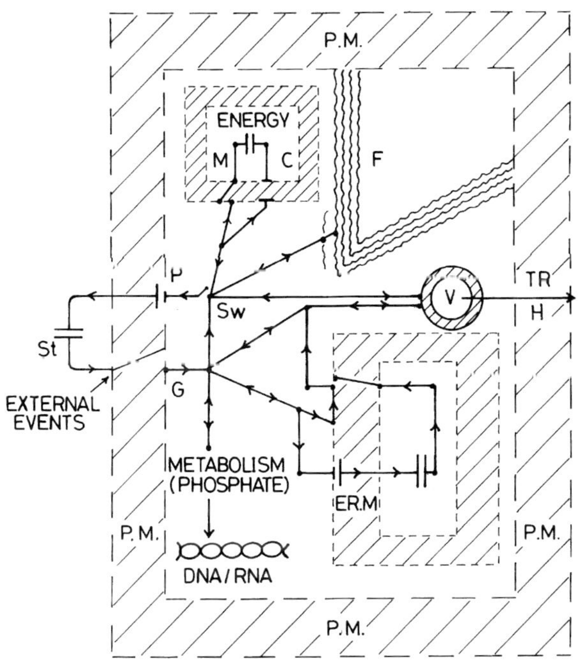

Fig. 14. Homeostasis illustrated as a circuit diagram in which a fixed current flows through the cell. The current here is of calcium ions. It travels to various "bits" in the circuit including the endoplasmic reticulum (ER), the mitochondria (M) and chloroplast (C), some vesicles or vacuoles (V) and passes through the plasma membrane (PM). The connection to DNA/RNA leads to protein *turnover*. *P* pump; *St* external store acting as a battery, *TR/H* transmitter or hormone release to maintain body homeostasis; *F* filament; S_W switch; *G* internal messenger, e.g. IP_3. The lines with arrows show the calcium currents

degree homeostasis is replaced by constant adjustment around limits since in higher animals cells constantly alter their activity. As stated early in the chapter, the evolution of the functions of this mineral element closely follows that of other signalling and control systems of connected inorganic elements. Hence, Table 11 indicates the evolution of Na^{+}/K^{+} ATP-ases and the increase in dependence of gene transcription on zinc. The calcium network depicted in Fig. 6 cannot stand alone in maintaining homeostasis but, by the very nature of the cooperative chemistry of living systems, must be linked to the activities of many other elements (Figs. 12 and 14). We need a view of cellular activity which is holistic as well as a study by reductive analysis.

References

1. Williams RJP (1998) In: Carafoli E, Klee C (eds) Calcium as a cellular regulator. Oxford University Press, New York, p 3
2. Dolmetsch RE, Lewis RS, Goodnow CC, Healy JI (1997) Nature 386:855
3. Chin ER, Olson EN, Richardson JA, Yang Q, Humphries C, Shelton JM, Wu H, Zhu W, Bassel-Duby R, Williams RS (1998) Genes and Development 12:2499
4. Williams RJP, Frausto da Silva JJR (1996) The natural selection of the chemical elements. Oxford University Press, Oxford
5. Neidhardt FC et al. (eds) (1996) *Escherichia coli* and *Salmonella*, 2nd edn. American Society of Microbiology Press, Washington, USA
6. Tisa L, Adler J (1992) Proc Natl Acad Sci USA 89:11804
7. Reusch RN, Huang R, Bramble LL (1995) Biophys J 69:754
8. Jones HE, Holland IB, Baker HL, Campbell AK (1999) Cell Calcium 25:265
9. Silver S (1996) In: Neidhardt FC et al. (eds) *Escherichia coli* and *Salmonella*, 2nd edn. American Society of Microbiology Press, Washington, USA, p 1091
10. Mannella CA, Marks M, Buttle K (1996) Trends Biochem Sci 116:93; Perkins G, Reuben C, Martone ME, Young ST, Ellisman M (1997) J Struct Biol 119:260
11. Williams RJP (1999) FEBS Lett (in press)
12. Means AR, Kahl CR, Crenshaw DGH, Dayton JS (1998) In: Carafoli E, Klee C (eds) Calcium as a cellular regulator. Oxford University Press, New York, p 512
13. Goodwin BC, Trainor LEH (1985) J Theoret Biol 117:79
14. Harrison LG (1993) Kinetic theory of living pattern. Cambridge University Press, Cambridge, UK
15. Evans DA, Williams LE (1998) Biochim Biophys Acta 1376:1
16. Trewavas AJ, Knight MR (1994) Plant Molec Biol 26:1329
17. Berridge MJ (1997) J Exp Biol 200:315
18. Gwathmey JK, Bentiregna LA, Ransil BJ, Grossman W, Morgan JP (1993) Cardiovasc Res 27:199
19. Williams RJP (1970) Q Rev Chem Soc (London) 24:331; J Bioenergetics 1:215
20. Tasaki I (1999) Japan J Physiol 49:125
21. McCormack JG, Halestrap AP, Denton RM (1990) Physiol Rev 70:391
22. Schäfer BW, Heizman CW (1996) Trends Biochem Sci 21:134
23. Soderling TR (1999) Trends Biochem Sci 24:232
24. Williams RJP (1998) Biochim Biophys Acta 1448:153
25. Stenflo J, Suttie JW (1997) Annu Rev Biochem 46:157
26. Johnson DR (1985) The Genetics of the skeleton. Oxford University Press, Oxford
27. Vincent JFV (1993) Biomechanics: materials. IRL Press, Oxford University, Oxford
28. Linse S, Johannson C, Brodin P, Grundstrom T, Drakenburg T, Forsén S (1991) Biochemistry 30:154
29. Skelton MJ, Kördel J, Akke M, Försen S, Chazin WH (1994) Nature Struct Biol 1:239
30. McConkey DJ, Orrenius S (1998) In: Carafoli E, Klee C (eds) Calcium as a cell regulator. Oxford University Press, New York, p 596

Evolution of EF-Hand Proteins

Susumu Nakayama[1], Hiroshi Kawasaki[2], Robert Kretsinger[3,*]

[1] Department of Biochemistry, Nagasaki University School of Medicine, 1-12-4 Sakamoto, Nagasaki 852-8523, Japan (*E-mail: f0539@cc.nagasaki-u.ac.jp*)
[2] Division of Plant Genetic Engineering, Kihara Institute for Biological Research, Yokohama City University, Maioka-cho 641-12, Totsuka-ku, Yokohama 244-0813, Japan (*kawasaki@yokohama-cu.ac.jp*)
[3] Department of Biology, University of Virginia, Charlottesville, VA 22901, USA (*rhk5i@virginia.edu*)

EF-hand proteins are homologous; they have all evolved from a common precursor. They contain from two to twelve copies of the EF-hand domain. The EF-hand is about thirty amino acids long and consists of an α-helix (*E*), loop, and second α-helix (F). Usually a Ca^{2+} ion is bound in the loop under physiological conditions; however, 30% of all known EF-hands do not bind calcium. With only a few exceptions all EF-hand-containing proteins are found in the cytosols of eukaryotic cells. Most of them function as transducers of information. In so-called quiescent cells the concentration of the free Ca^{2+} ion is about $10^{-7.2}$ M; following a stimulus to the cell $[Ca^{2+}]$ rises to $\sim 10^{-5.8}$ M. Many EF-hand proteins are calcium modulated. In a quiescent cell they are in the apo or in the magnesium form; following stimulation they are in the calcium form. These calcium signals are tuned by complex spatial and temporal distributions, as well as by differing affinities and binding rates of calcium-modulated proteins. The focus of this review is to summarize the characteristics of the 66 known subfamilies of EF-hand proteins and to suggest general patterns of their evolution.

Keywords. Calcium coordination, Cell signaling, Calmodulin, Dendrogram, Evolution

1 Introduction

Although few contemporary biochemists would admit to an interest in classification, especially to their funding agencies, the ordering of objects was and remains the foundation of most natural sciences. First, one is obliged to examine the appropriate criteria for classification. Second, an appropriate grouping of elements, or of stars, or of proteins should suggest, if not predict, properties of the objects being classified. Conversely, their properties might be rationalized by the classification. In biology classification also lends insight into the process of evolution. To these ends, rationalization of properties and insight into evolution, we summarize our classification of proteins that contain EF-hand domains.

The EF-hand domain was first recognized in the crystal structure of parvalbumin and it was proposed that it had evolved by the process of gene triplication and fusion [1]. To date the amino acid sequences of 581 proteins that contain 2073 EF-hand domains have been published or deposited in data banks. We have classified these into 66 distinct subfamilies that contain from two to twelve copies of the EF-hand. Crystal structures are available for fifteen of these subfamilies.

The main criterion for classifying these proteins has been similarity in amino acid sequence. Since the EF-hand domain is only 30 residues long, the statistical significance of standard amino acid alignments using programs like FASTA [2] is weak. Hence, we have used four additional criteria both to identify EF-hands and subsequently to place them in subfamilies.

Members of the same subfamily have similar characteristics and functions. This division is sometimes difficult for proteins that are congruent (as described later), for example, intestinal calcium-binding protein (ICBP) vs. S100 or troponin C (TNC) vs. troponin from nonvertebrates (TPNV). When trying to identify distantly related EF-hands, we have placed additional weight on certain residues in alignments of amino acid sequences, for instance, those involved in calcium coordination. Here one must be especially cautious because some EF-hands bind a calcium ion (Ca^{2+}) and some do not. Calcium binding is not necessarily a distinguishing characteristic, since in some subfamilies the same (aligned) EF-hand does or does not bind Ca^{2+} depending on the species of origin. Further, some EF-hands bind Ca^{2+} with alternate ligands, note a, b, c and d in Table 1 and Fig. 1. The third criterion, also not absolute, is the observation that EF-hands usually occur in pairs, adjacent in sequence. This "guilt by association" strengthens the interpretation of homology for the second EF-hand of aequorin (AEQ) and the fourth EF-hand of the nematode calcium-binding protein (EF12). Finally all members of a subfamily must, within the error of analysis, be congruent with one another. Members of different subfamilies may, or may not,

Table 1. Summary of characteristics of 66 subfamilies

	Name	APFP	Funct/Struct	1 (7)	2 (8)	3 (9)	4 (10)	5 (11)	6 (12)
CTEF									
CAM	calmodulin	APFP	+X	+	+	+	+/−		
TNC	troponin C	A...	+X	+/−	+	+/−	+		
ELC	essential light chain, myosin	A.F.	+X	a/−	+/−	+/−	+/−		
RLC	regulatory light chain	A.FP	+X	+	−	−	−		
TPNV	troponin, nonvertebrate	A...	+?	−	+	+/−	+		
CLAT	CAM-like leaf (*Arabidopsis*)	P..	??	+	+	+	+***		
SQUD	squidulin (*Loligo*)	A...	+?	+	+	+	+		
CDC	CDC31 and caltractin	APFP	??	+	+/−	+/−	+		
CAL	cal1 (*Caenorhabditis*)	A...	??	+	+	+	+		
CAST	CAST	.P..	??	***+	−	+	+		
CPV									
CLNB	calcineurin B	A.F.	+X	+	−	+	+		
P22	p22	A...	??	+	−	+	+		
VIS	visinin and recoverin	A...	+X	−	+	+	−		
CALS	calsenilin (*Homo*)	A...	??	***−	−	+	+		
DREM	DRE antag. modul (*Homo*)	A...	+?	***−/?	+/?	+	+		
CMPK	CAM dep prot kinase (*Lilium*)	.P..	+?	***+	+	+	+		
SOS3	Ca sens homo (*Arabidopsis*)	.P..	+?	−	−	−	−		
Pairings									
RTC	reticulocalbin (*Mus*)	A...	??	+	?/+	?/+	+	+	+
SCF	DNA supercoil fact (*Bombyx*)	A...	??	***+	−	+	+	+	+
CALP	calpain	A...	+X	***+	+	−	−	−	
SORC	sorcin/grancyclin	A.F.	+?	***+	+	−	−	−	
S100	S100	A...	?X	b/−	+				
ICBP	intestinal Ca-binding protein	A...	?X	b/−	+				
HYFL	trichohylin profilag	A...	??	b/?	+***				
DGK	diacylglycerol kinase	A...	+?	***+	+***				
NUBN	nucleobindin and NEFA	A...	??	***+	+***				
CRGP	CAM rel gene product (*Homo*)	A...	??	+	+				

Table 1. (*Contd.*)

	Name	APFP	Funct/Struct	1 (7)	2 (8)	3 (9)	4 (10)	5 (11)	6 (12)
ACTN	α-actinin	A.F.	+?	***+/–	+/–				
FDRN	α-spectrin and α-fodrin	A...	+?	***–	–				
GPD	glycerol-P-dehydrogenase	A...	+?	***–	+				
AIF1	allograft inflammatory factor	A...	??	***+	+/?				
BM40	osteonectin, SPARK	A...	?X	***c	+				
QR1	QR1 and SC1	A...	??	–	+				
Self									
EF12	Ca-binding prot of nematodes	A...	??	***+	+	+	–	+	+
				+	+	+	+	+	+
LPS	*L. pictus* SPEC resembl prot	A...	??	+	+	+	+	+	+
				+	–				
CLBN	calbindin 28 kDa, calretinin	A...	??	+	+/–	+	+	+	–
EP15	EP15	A...	+X	***–	–	–***	+***	+	–***
TCBP	tetrahymena CaBP	...P	??	–	+	+	+		
P26	p260lf (*Rana*)	A...	??	–	–***	–	–		
PLC	phospholipase C	A.F.	+X	***–	–	–	–***		
CBP	CBP1, CBP2 (*Dictyostelium*)	..F.	??	+	+	+	+		
Miscellaneous									
PFS	surface protein (*Plasmodium*)	...P	??	***+	+	+	+	+	–***
CDPK	Ca-dependent protein kinase	.P..	+?	***+	+	+	+		
PFPK	protein kinase (*Plasmodium*)	...P	+?	***+	+	+	+		
SPEC	*Strongylocentrotus* CaBP	A...	??	+/–	+	+	+/–		
TPP	p24 thyroid protein	A...	??	+	+	+	?		
1F8	1F8 and TB17 and calflagin	...P	??	+	+	+	+		
SARC	sarcoplasm Ca-binding protein	A...	?X	+	+/–	+	+/–		
AEQ	aequorin and luciferin BP	A...	+?	+	*–*	+	+		
PPTS	protein phosphatase	A...	+?	***–	*–*	+	+		
H32	HRA32 (*Phaseolus*)	.P..	??	+	+	+	+		
EFH5	EFH5	...P	??	–	+/–	+/–	–		
CVP	Ca-vector prot (*Branch*)	A...	??	–	–	+	+		

PMAT	memb. assoc (*Arabidopsis*)	.P..	??	***+	+	+	+
LAV	LAV1 (*Physarum*)	..F.	??	***+	+	+	+
CMSE	CaBP (*Saccharopolyspora*)	bact	??	+	+	+	+
MSV	MSV097 (*Entomopoxvirinae*)	virus	??	+	+	–	–
PARV	parvalbumin	A...	?X	del	–	+	+
FIME	fimbrin	A.F.	+?	+/–	+/–***		
GRP	ras guan releasing protein (*Rattus*)	A...	??	***+/?	+/?***		
PKD	PKD2L/polycystin	A...	??	***–	+***		
RYR	ryanodine receptor/Ca release	A...	+?	***–	+***		
CBL	proto-oncogene Cbl	A...	+X	***–	d***		
CIB	Ca- and integrin-binding protein	A...	??	***+	+		
SENS	calsensin (*Haemopsis*)	A...	??	+	+		
GRV	groovin (*Drosophila*)	A...	??	+	+		

The EF-hand subfamilies are described in several groups:
CTER, those that are congruent with *c*almodulin, *t*roponin C, *e*ssential, and *r*egulatory light chains;
CPV, those that are congruent with *c*alcineurin B, *p*22, and *v*isinin (recoverin);
Pairings, those closely related between or among themselves but not closely related to other subfamilies;
Self, those (some of) whose EF-hands are most closely related to other EF-hands within the same subfamily; and
Miscellaneous, those whose domains do not show a strong and consistent pattern of similarity with other EF-hand subfamilies. For 28 subfamilies only one sequence is available; this is indicated by inclusion of genus name in parentheses or as part of the name of that subfamily.
APFP refers to whether the protein is found in Animals, Plants, Fungi, and/or Protists.
The *Funct/Struct* columns indicate whether a function is known "+" or not "?" and whether a crystal structure is available "X" or not "?". The symbol "***" before the first EF-hand column or after the last column indicates a heterochimeric protein with the non-EF-hand domain(s) to either the N- or C-side of the EF-hands. For F26 "***" indicates another domain between EF-hands 1 and 2 and EF-hands 3 and 4. For AEQ and PPTS the "**-**" indicates that domain is not recognizable as an EF-hand by analysis of it sequence; however, its proximity to an otherwise unpaired EF-hand suggests that it may be an EF-hand and that it is inferred not to bind calcium. For each EF-hand is indicated whether calcium binding is observed (or inferred from sequence) "+" or not "–". For some subfamily EF-hands there are instances of both binding and not binding calcium "+/–". There are four examples – "a, b, c and d" – of non-canonical EF-hand loops that bind calcium. Some loops inferred not to bind calcium may provide additional examples of non-canonical calcium-binding loops. EF12 has twelve and LPS has eight EF-hands; for ease of formatting, EF-hands 7–12 are listed under EF-hands 1–6. The NMR structures of the first of three EH-domains (Whitehead B, Tessari M, Carotenuto A, van Bergen en Henegouwen PMP, Vuister (1999) Biochemistry 38:11271) and of EH2 (de Beer T, Carter RE, Lobel-Rice KE, Sorkin A, Overduin M (1998) Science 281:1357) from Eps15 both reveal a pair of EF-hands.

be congruent with one another; if congruent, these subfamilies are referred to as a "group", e.g. *CTER* and *CPV*. Congruence, in this context, is best illustrated by example, to be elaborated later. If the individual proteins, e.g. TNCs, within a subfamily are divided into N-halves and C-halves, then the dendrogram computed from the N-halves should be the same as the dendrogram from the C-halves. These four criteria – function, Ca^{2+} binding, pairing of domains, and congruence – in addition to sequence alignment, are noted in the descriptions of the 66 subfamilies and their relationships to one another.

In accord with this approach to classification and analysis we first address several patterns of function. Next we summarize a few of the structural characteristics involved in Ca^{2+} binding and in the formation of pairs of EF-hands. Finally we illustrate congruence and its discriminatory power.

The main part of this review consists of a tabulation of the 66 subfamilies with brief comments about the distinguishing features of each. We conclude, returning to our title, with interpretations of the evolution of proteins that contain EF-hands.

2 Characteristics of EF-Hands and EF-Hand Proteins

2.1 Functions of EF-Hand Proteins

In the most simple scheme a calcium-modulated protein is in the apo form in a quiescent cell with $[Ca^{2+}]$ $\sim 10^{-7.2}$ M. Following cell stimulation, $[Ca^{2+}]$ rises to $\sim 10^{-5.8}$ M and the protein binds a Ca^{2+} ion in its EF-hand(s). This induces a change in conformation enabling the EF-hand protein to bind and activate a target enzyme. There are many known, and certainly more yet to be discovered, variations or elaborations on this scheme.

1. By far the most egregious approximation is the simple model of pCa_{out} $\sim$2.8, pCa_{cyt} $\sim$7.2 in a quiescent cell, then a stimulus with $[Ca^{2+}]$rising to pCa_{cyt} $\sim$5.8, then falling to $\sim$7.2 with relaxation. Many cells maintain a tonic activity – seldom maximally stimulated or fully quiescent. Further, $[Ca^{2+}]$ varies across the cytosol with distance and time. Of special relevance, there are constant leaks of calcium into the cytosol through the membranes of the cell surface and of the endoplasmic reticulum balanced by ever active calcium extrusion pumps. This results in the $[Ca^{2+}]$ being higher near these membranes than the average value in the cytosol.
2. Although we speak of apo and calcium forms, the situation is more complex. In most EF-hand domains the affinity for calcium is about 10^4 times stronger for Ca^{2+} than for Mg^{2+}. With $[Mg^{2+}]$ in the cytosol near constant, pMg $\sim$2.8, this EF-hand domain would be primarily in the apo form in the quiescent cell. In contrast, if the $pK_d(Ca^{2+})$ is $\sim$6.8 and $pK_d(Mg^{2+})$ $\sim$2.8, then this EF-hand would be at least half magnesium in the resting cell. At higher affinity for divalent cations, $pK_d(Ca^{2+})$ $\sim$7.4 and $pK_d(Mg^{2+})$ $\sim$3.4, the EF-hand would be partially in calcium and partially in magnesium forms in the resting cell. For instance, EF-hands 3 and 4 of troponin C (TNC) appear to be mostly in the calcium form in the resting

Canonical

```
                        ------------- Loop ---------------
 αααααααααα E αααααααα                       αααααααα F ααααααααααα
                    1 1 1 1 1 1 1 1 1 1 2  2 2 2 2 2 2 2 2 2
 1 2 3 4 5 6 7 8 9  0 1 2 3 4 5 6 7  8 9 0  1 2 3 4 5 6 7 8 9
 E n * * n n * * n  X * Y * Z G # I -X * * -Z n * * n n * * (n)
```

ELC, domain **1**

```
                                     1 1 1 2 2 2 2 2 2 2 2 2 2
 1 2 3 4 5 6 7 8 9                   7 8 9 0 1 2 3 4 5 6 7 8 9
 $ n * * V F $ L F  D F W D G R D G * V D A n K n G D n n R C L
                    X  -X  -Y
                    Y   Z       -Z
                    =O  =O      =O
```

S-100, domain **1**

```
                    -8   -5   -3   0          +5
                                              2 2 2 2 2 2 2 2 2
 1 2 3 4 5 6 7 8 9                            1 2 3 4 5 6 7 8 9
 * n * * n n * * n  * *  *E G D * * * L * K * E n * * n n * * *
                    X     Y  Z  -Y            -Z
                    =O    =O =O =O
```

BM40, domain **1**

```
                                1 1 1 1 1 1 2  2 2 2 2 2 2 2 2 2
 1 2 3 4 5 6 7 8 9 0 1  2 3     4 5 6 7 8 9 0  1 2 3 4 5 6 7 8 9
 P n * W * n * * L D *H~P *     D * # L -X * * -Z n * P n * * P n
                        =O
```

Fig. 1. Consensus sequences of EF-hands and Ca^{2+}-binding ligands

muscle cell and serve a structural function while EF-hands 1 and 2 are mostly apo and, following stimulation, function to transduce information. The actual functions of calcium-modulated proteins are exquisitely dependent on their $pK_d(Ca^{2+})$s and $pK_d(Mg^{2+})$s and on the spatial and temporal distributions of the calcium pulse. This fine tuning is not captured in simple on/off diagrams.

3. The targets of the EF-hand, calcium-modulated protein need not be an enzyme. Proto-oncogene Cbl protein (CBL) binds to DNA. α-Actinin (ACTN), fordrin (FDRN) and fimbrin (FIMB) interact with thin filaments. Trichohyalin (HYFL) interacts with keratin-based intermediate filaments
4. A single EF-hand protein may interact with multiple targets. Calmodulin (CAM) regulates over 20 enzymes or structural proteins.
5. An EF-hand protein may be free in the cytosol in the apo form and associate with its target in the calcium form. Or the EF-hand protein, such as calcineurin B

(CLNB), may bind to its target calcineurin A (a protein phosphatase) in both apo and calcium forms.

6. This permanent attachment to a target may be taken one step further. Thirty-one of the 66 subfamilies are heterchimeras; the gene encoding EF-hands has fused with the gene encoding other domains. Of these 31, functions are known for only fourteen (four kinases, a phosphatase, lipase, dehydrogenase, protease, three that interact with the cytoskeleton, and one each that bind to a ryanodine receptor, repressor, and DNA).
7. All of the cytosolic EF-hand proteins are inferred to be calcium modulated. However, they need not be directly involved in information transduction. Several appear to be involved in temporal buffering of calcium, such as parvalbumin (PARV), and/or calcium transport, ICBP. In a resting muscle cell PARV, with relatively high affinity for divalent cations, is (primarily) in the magnesium form. EF-hands 1 and 2 of TNC have low affinity for divalent cations and are primarily apo. Following stimulation the influx of calcium binds *first* to TNC, which has a *lower* affinity for calcium than does PARV, because it takes a few milliseconds for the Mg^{2+} to dissociate from the PARV. The Ca^{2+} then diffuses to the apo sites on the PARV thereby facilitating and sharpening the relaxation process and queuing the Ca^{2+} for the calcium extrusion pump [3].
8. The preceding overview of functions assumes the existence of EF-hands as calcium-modulated proteins in the cytosol. Reticulocalbin (RTC) has a leader sequence typical of proteins found within the lumen of the endoplasmic reticulum and glycerol phosphate dehydrogenase (GPD) is found on the outer surface of the inner mitochondrial membrane. Osteonectin (BM40) and QR1 are extracellular in an environment whose pCa is near constant $\sim$2.9. One EF-hand of BM40 binds calcium, supposedly with high affinity, but not with any modulation. This appears to be an example of Nature having taken a protein initially "designed" for one function and put it to use in extracellular stabilization. S100 and PARV are sometimes found extracellularly. Whether this reflects a normal function or pathology has yet to be determined.
9. Although bacteria extrude calcium to maintain pCa $\sim$7.0 in the cytosol, there is no evidence of their using calcium as a cytosolic messenger. There is one example of an EF-hand protein (CMSE) in a prokaryote, *Saccharopolyspora*, perhaps transduced from a eukaryote. An EF-hand protein (MSV) is encoded by a virus, *Entomopoxvirinae*.

A great deal of variation in structure is realized from this seemingly simple, 30-residue domain. When convoluted with the finely shaped pulse(s) of messenger calcium, a rich tapestry of information can be transduced by EF-hand proteins.

2.2
EF-Hands and Ca^{2+} Binding

Most functions of EF-hand proteins involve, or are inferred to involve, calcium binding. However, 81 of the 247 domains of the 66 subfamilies do not bind calcium. In some domain subfamilies, indicated in Table 1 by +/−, some representatives do and some do not bind calcium. For a few, one is reluctant to predict, e.g. ?/+. Four subfamily domains have non-canonical Ca^{2+} coordination, indicated by a, b, c, or d in Table 1 and illustrated in Fig. 1. The important point is that one should not assume that all EF-hand domains bind calcium nor that one can always predict calcium binding, let alone affinity, from the amino acid sequence.

The consensus sequence shown in Fig. 1 is a valuable heuristic for illustrating and understanding the main characteristics of an EF-hand. We emphasize, however, that the most sensitive search involves testing a candidate protein against a large database, not against the heuristic. Such standard searches involve several complications. One EF-hand is so short that an heterochimeric protein will find homologs with higher significances based its non-EF-hand regions. The portions of the test sequence that are identified as similar to other proteins should be deleted and the remaining sequence subject to a new search. A second caution is that the proteins in the database that are most similar to the test sequence (or to its remainder) may not be indicated in the database descriptor as containing EF-hands, hence the value of Table 1. Conversely, one can search the database including the test sequence of interest with a known EF-hand. Again because one EF-hand is so short and because there is such a range of EF-hand sequences, one is well advised to search with several disparate EF-hands and/or to search with pairs of EF-hands. It is important to compare the significance of any alignment with the distribution of alignment scores from the entire database.

The canonical EF-hand consists of α-helix E, (residues 1–10), a loop around the Ca^{2+} ion (10–21) and α-helix F (19–29). Residue 1 is often Glu ("E"); the insides of the helices usually have hydrophobic residues "n". The side chains of five residues, approximated by the vertices of an octahedron (X, Y, Z, -X and -Z), provide oxygen atoms to coordinate Ca^{2+}; residue 16, "#" at -Y, bonds to Ca^{2+} with its carbonyl oxygen. The calcium is actually seven coordinate in a pentagonal bipyramid with X, -X major axis. There are five oxygen atoms in the Y, Z plane; since the -Z ligand, usually Glu, coordinates Ca^{2+} with both oxygen atoms of its carboxylate group. A Gly, "G", at 15 permits a tight bend; "I" at 17 has a hydrophobic side chain that attaches the loop to the hydrophobic core of the pair of EF-hand domains.

Four variations to this canonical calcium coordination scheme have been confirmed in crystal structures. (1) Loop 1 of molluscan ELC has two residues inserted. Asp10 coordinates calcium or magnesium with both carboxylate oxygen (X) and carbonyl oxygen (Y); Asp12 (Trp is inserted between 11 and 12) coordinates with both carboxylate (-X) and carbonyl oxygen (Z); Asp14 (Arg is inserted between 13 and 14) uses its carboxylate (-Y); and a carbonyl oxygen comes from 16 (-Z). The "$...$" in the molluscan ELCs is either Glu...Asp or Asp...Glu. (2) In S100 there are two residues inserted into the first loops of those domains 1 that bind calcium but not at homologous positions to those inserted in loop 1 of ELC. The Glu at -Z coordinates with both oxygens of its carboxylate group. Four carbonyl oxygens (X, Y, Z, and -Y) coordinate calcium. The seventh ligand is the oxygen of water. (3) Loop 1 of BM40 is similar to the canonical loop except that a Pro is inserted between 12 and 13. The His–Pro peptide bond is cis; the result is that the carbonyl oxygen of Pro bonds to calcium at Y instead of the usual side chain of residue 12. Pro residues are at 24 and 28. For all three variants conserved or consensus residues that differ from canonical are indicated. (4) The distributions of ligands in the second EF-hand of CBL, the fourth variant, is canonical (Asp, Thr, Asn, Ser, Glu); however, Ser247 does not coordinate Ca^{2+}, instead the -X ligand is provided by Glu164 of the fourth helix of a preceding four-helix domain.

The canonical calcium-binding loop of the EF-hand is best regarded as a reference point for evaluating the numerous variations. Positions X (residue 10), Y (12), Z (14), -Y (16), -X (18), and -Z (21) represent the vertices of an octahedron. However, the Ca^{2+} ion is seven coordinate in a pentagonal bipyramid, with major axis along X. Six amino acids are involved in Ca^{2+} coordination. Since both oxygen atoms of the carboxylate group at -Z bind, the coordination number is seven. Usually Asp or Asn is found at X and Y; Asp, Asn, or Ser at Z; the carbonyl oxygen of a variety of residues is a -Y; -X is more variable but usually Asp, Asn, or Ser; and usually Glu at -Z. Although one can produce a scheme that correlates calcium affinity with distribution of residues, we know of no scheme that predicts affinities for EF-hand loops that were not included in making the scheme. This is because the free energy of the system depends on the (change in) conformation of the entire protein, not just the loop in question.

Little is known of the coordination of the Mg^{2+} ion in EF-hand loops. However, given the many precedents [4] from the structures of small molecules, one can safely infer that the Mg^{2+} ion will be six coordinate. Probably the bidentate carboxylate at -Z rotates to become monodentate and the other oxygen atoms are ~0.2 Å closer to the Mg^{2+} than to the Ca^{2+}.

Three (a, b, and c in Table 1) of the non-canonical calcium coordinations involve several carbonyl oxygen atoms in place of the usual oxygen-containing side chains. None of these coordinations were predicted; hence the caution when considering the prediction (without crystal structure) of non-binding, "-", in Table 1. The fourth exception, d in Table 1 for CBL, is the use of the side chain of a Glu from a nearby α-helix instead of the Ser at the -X vertex in the second EF-hand loop, which by sequence appears to be canonical.

The angle between helix E and helix F varies between apo and calcium bound. This is important because the targets, discussed in the next two sections, of these calcium-modulated proteins, like CAM, are α-helices that fit into the groove between the two EF-hands of a pair. The accessibility and hydrophobicity of this groove is critically dependent on the interhelical angles of the two EF-hands. To first approximation the interhelical angle, see Fig. 2 and legend, is closed in the apo form and more open in the calcium form. However, we emphasize four points: The angles associated with closed forms vary over a broad range as do the angles of open conformations. Second, the change in orientation associated with calcium binding varies. Third, several EF-hands of ELC and of RLC do not bind calcium. Their conformations are invariant to $[Ca^{2+}]$ but important to their functions. They cannot be simply classified as open or closed [5]. Finally, many of these EF-hands are, as will be discussed, in the magnesium, not the apo form in the quiescent cell; we have little information about the conformations of magnesium EF-hands.

2.3 Pairs of EF-Hands

With only a few exceptions EF-hands occur in adjacent pairs. The fifth EF-hands of m and μ calpains (CALP), and probably those of the close homolog sorcin (SORC), pair to form a dimer. The N-terminal domain of PARV, indicated as #2 in Table 1, covers the hydrophobic surface of the 3,4 pair.

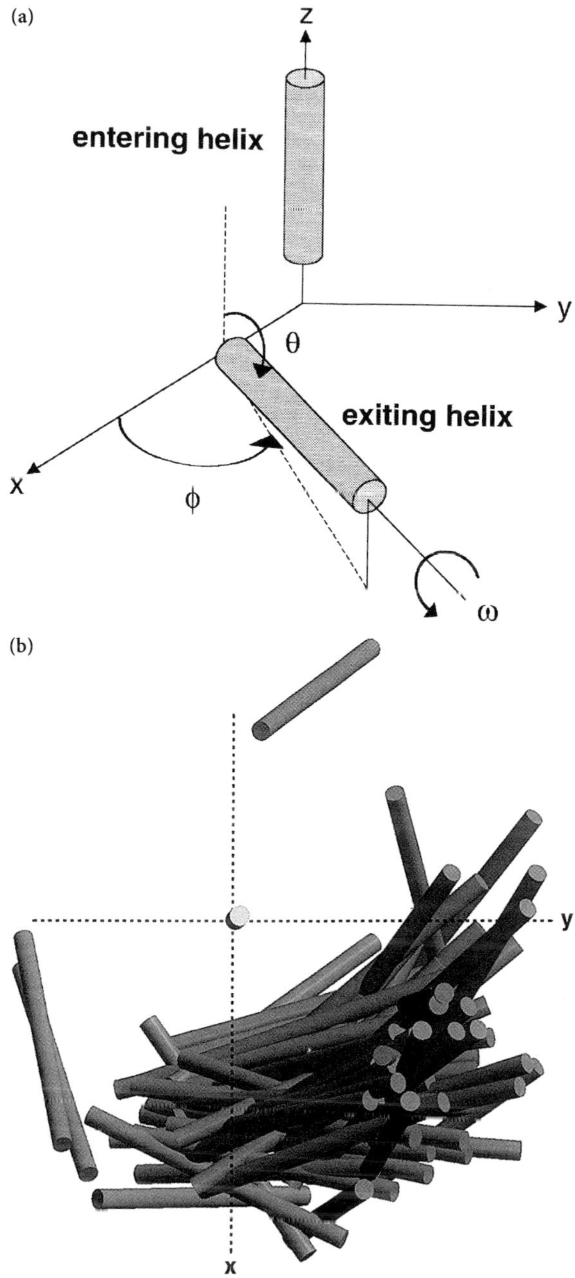

Fig. 2. Distribution of angles relating helix E and helix F [5]. **a** Helix E is superimposed on a reference EF-hand on the Z-axis. The orientation of helix F is evaluated using the angles θ, φ, and ω. **b** Superposition of all EF-hands of known structure

Just as there is a large range of angles between helix E and helix F of a single EF-hand so there is also a range of relationship between the first (*ODD* numbered) and second (*EVEN*) EF-hand. Both, either, or neither might bind calcium.

When both bind calcium they may do so cooperatively but this is still subject to debate.

There are three general patterns among the fourteen subfamilies of known structure. The EF-hand protein is involved in information transduction and an α-helix of the target protein lies in the hydrophobic groove between two EF-hands. This groove in CAM (Fig. 3) and in TNC is more exposed when the two EF-hands are in the calcium forms and their respective helices E and F are more open. In the second pattern an α-helix of a chimeric protein lies within the *ODD*, *EVEN* groove of the same protein. This self binding is inferred to be, at least in some instances, a component of the information transduction pathway. In the third pattern the EF-hand protein is involved in calcium buffering or transport and the *ODD*, *EVEN* groove is either covered, as in PARV by EF-hand 2, or partially occluded by the 1,2 loop, as in ICBP. One can safely anticipate that other patterns of function will be revealed as more structures are determined.

2.4
Congruence Within and Among EF-Hand Subfamilies

Congruence is an important characteristic in assigning proteins to subfamilies and in grouping subfamilies such as CTER and CPV. By definition all members of a

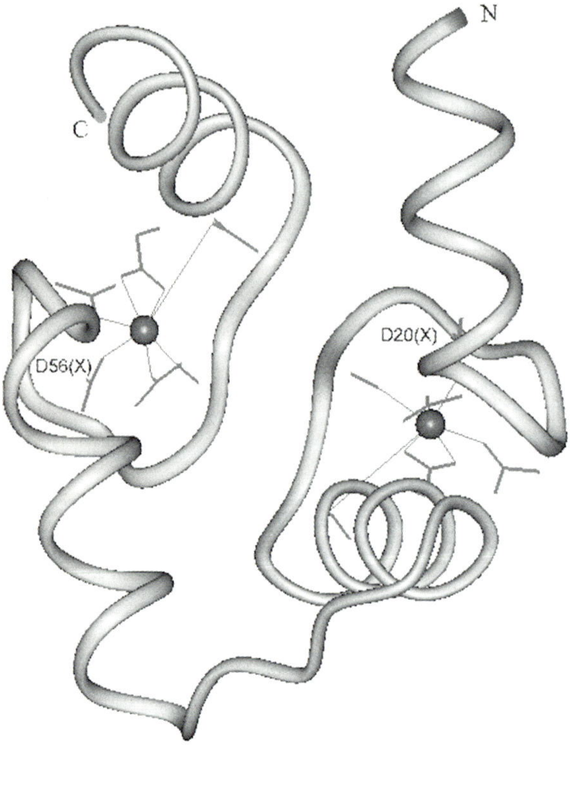

Fig. 3. Pair of EF-hands. The main chain of EF-hands 1 and 2 of CAM are viewed down their approximate two-fold axis from the hydrophobic core

subfamily must be congruent, as illustrated in the dendrogram computed from the 56 domains of fourteen TNCs from ten different species (Fig. 4). The dendrogram illustrates the two essential characteristics of congruence. All domains 1 are more closely related to one another than to other domains; correspondingly all domains 2 cluster together as do domains 3 and 4. Second, the distribution of domains within each of the dendrograms of the four subdomains is (nearly) identical. An additional interesting characteristic, not inherent to the concept of congruence, is that the cluster of domains 1 is most closely related to domains 3 and the domains 2 are more closely related to domains 4. This reflects a gene duplication and fusion of an *ODD*, *EVEN* pair of EF-hands in the ancestor of all animals that have TNC.

Correspondingly, if two subfamilies are congruent, the dendrograms computed from their constituent proteins will be (nearly) identical to those computed from the corresponding domains of those constituent proteins. For instance all ten subfamilies within CTER are inferred to have evolved from a common, four-domain precursor by gene duplication and subsequent divergent evolution (Fig. 5).

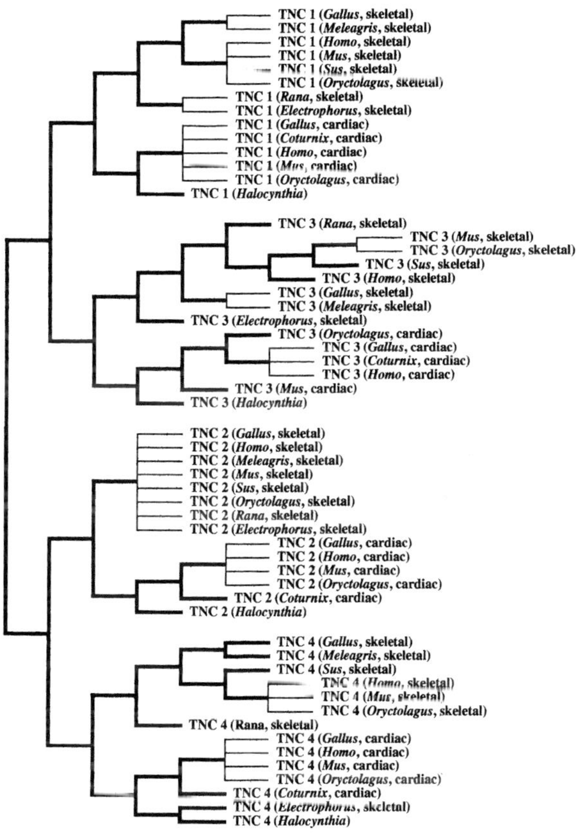

Fig. 4. Congruence within the troponin C subfamily. The dendrogram was computed as described in the text

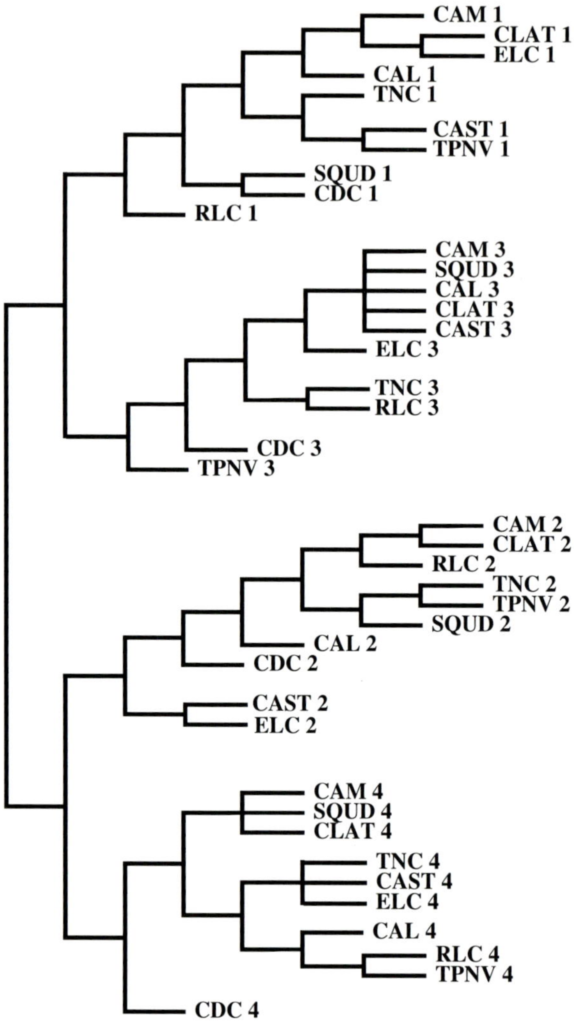

Fig. 5. Dendrogram of CTER. The ten subfamilies – *CAM*, *TNC*, *ELC*, *RLC*, CLAT, SQUD, CDC, TPNV, CAL and CAST – show reasonable, but imperfect, congruence. All EF-hands 1 cluster together, as do EF-hands 3, 2, and 4. Branch lengths and deviations are not shown; EF-hands 3 are the most divergent

3 Subfamilies of EF-Hand Proteins

3.1 CTER Group

Calmodulin (CAM). Calmodulin (CAM) is inferred to be encoded in all eukaryotes and expressed in most cells. In its four-calcium form it activates over 20 target enzymes or structural proteins [6]. Pairs of domains 1 and 2 are connected to lobe 3,4 by an eight-residue linker. Helix F2, the linker, and helix E3 form a seven-turn α-helix. The linker serves as a flexible tether. In the apo form both lobes are in the closed conformation and the tether is extended; CAM is then dumbbell shaped. In the Ca_4-form the tether is bent, both lobes are in an open conformation, and the two lobes enfold a helix of the target protein [7]. The ten subfamilies (CAM, *T*NC, *E*LC, *R*LC TPNV, CLAT, SQUD, CDC, CAL and CAST) are congruent (Fig. 5).

Troponin C (TNC). Troponin C (TNC) together with TNI and TNT form the heterotrimer, troponin. It imparts calcium sensitivity to muscle. TNC has 11 residues in the linker; hence, the F2 α-helix, linker, E3 α-helix is eight turns long. TNC is dumbbell shaped in the apo form. The two lobes of TNC in the Ca_4-form enfold the target peptide of TNI, residues 1–47 [8], in a conformation similar to the CAM* target complex. Most animals have several isoforms encoded at separate loci. In the dendrograms based on both protein and DNA, the vertebrate skeletal and cardiac forms cluster separately indicating that the common precursor of all vertebrates already has these two isoforms. *Halocynthia*, a protochordate, has the most divergent TNC; its TNC clusters near the base of the cardiac branch (Fig. 4).

Essential Light Chain (ELC). Essential light chain (ELC) has two alternate mRNA splice forms. A1 (long) has ~55 residues before the first EF-hand; A2 (short) has about ten. The ancestor to vertebrates had four isoforms of ELC. The skeletal form is most divergent, then the smooth muscle form. The cardiac atrial and cardiac ventricular forms are most similar. Molluscan ELC is inferred to bind calcium in its first EF-hand; vertebrate ELCs do not bind calcium [9].

Regulatory Light Chain (RLC). Regulatory light chain (RLC) of vertebrate smooth and cardiac myosins do not bind calcium but are an alternate source of calcium regulation via a CAM-activated kinase. Comparison of the crystal structure of molluscan RLC with a Ca^{2+} ion bound in EF-hand 1 with both chicken ELC and RLC complexed with the S1 fragment of myosin permitted Houdusse et al. [10] to present a model of ELC and RLC complexed with myosin in both active and inactive states.

Troponin, Nonvertebrate (TPNV). Troponin, nonvertebrate (TPNV) is found in various nonvertebrates; for example, lobster has three isoforms [11]. It is a close homolog of TNC; however, its mode of function remains unknown and it clusters in a subfamily distinct from TNC. The interdomain linkers of CAM, TNC, ELC and RLC are bent when those proteins are bound to a target. In contrast, the 2,3 linker of TPNV is predicted to be strongly helical and hence not to bend.

Calmodulin-like Leaf Protein from *Arabidopsis thaliana* (*CLAT*). Calmodulin-like leaf protein from *Arabidopsis thaliana* (CLAT) is congruent with *CTER* but differs from CAM in that it has a 47-residue domain of unknown homology at its C-terminus [12]. Its linker between lobe 1,2 and lobe 3,4 is TYSEK in contrast to the MKDTDSEE of CAM.

Squidulin (SQUD). Squidulin (SQUD) is identical to vertebrate CAM in 101 of 149 positions; however, its Lys115 is not methylated. Lys115 of CAM is trimethylated. Further, it has a Pro at the second residue of the 2,3 linker, which is helical in CAM when it is not bound to a target. SQUD co-purifies with the 235 kDa myosin of squid axoplasm and is inferred to function as a myosin light chain [13].

Caltractin (CDC). Caltractin (CDC) was initially cloned from the unicellular green alga, *Chlamydomonas*. It is a structural component of the basal body complex, the major microtubule organizing center in *Chlamydomonas* and the functional homolog of the centrosome in the animal cell. The *CDC31* gene product, a yeast homolog of caltractin, interacts with the *KAR1* product. *KAR1* encodes an essential component of the yeast spindle pole body that is required for karyogamy and duplication of the spindle pole body [14].

Cal1 (CAL). The *cal-1* gene was isolated from the nematode *Caenorhabditis elegans* by hybridization with a human CAM cDNA probe [15]. CAL has four EF-hands in its 161 residues. It is distinct from CAM but congruent with *CTER*.

pCAST **(CAST).** The *pCAST* (CAST) clone from *Solanum tuberosum* (potato) encodes four EF-hands preceded by 38 residues [16]. Its 2,3 linker is 23 residues long. It is not in the same subfamily as PMAT from *Arabidopsis*, which is not congruent with CTER; however, it is interesting to note that both contain $(Ser)_5$ in their N-terminal domains.

3.2 CPV Group

Calcineurin B (CLNB). Calcineurin B (CLNB) is the regulatory subunit and calcineurin A is the catalytic subunit of a protein phosphatase. Calmodulin also binds to calcineurin A; both CLNB and CAM in their calcium forms are required for catalytic activity. The crystal structure of Ca_4-CLNB, complexed with an α-helical fragment (350–370) of calcineurin A and with the drug complex FKBP12-FK506 [17, 18], shows both lobe 1,2 and lobe 3,4 embracing the fragment in their grooves homologous to those of CAM and TNC. In CLNB the two lobes bind to the same side of the target and are displaced about 22 Å along it. In contrast, the two lobes of CAM enfold the target peptide from myosin light chain kinase on opposite sides and are related by an approximate two-fold rotation axis. The seven subfamilies (CLNB, P22, VIS CALS, DREM, CMPK and SOS3) are reasonably congruent (Fig. 6).

P22 (p22). P22 (p22) is required for constitutive, exocytotic membrane traffic in rat liver. p22 is N-myristoylated, as is recoverin. The N-myristoylate group and the calcium-induced change in conformation are essential to the function of P22 [19]. Lin and Barber [20] reported that P22 inhibits the GTPase-stimulated Na^+/H^+ exchanger.

Recoverin or Visinin (VIS). Recoverin or visinin (VIS) regulates guanylate cyclase activity in retinal rod outer segments of vertebrates. In the dark, cationic channels are kept open by bound cGMP. Ca^{2+} enters through these channels; this influx is matched by efflux through a Na^+-K^+, Ca^{2+} exchanger. Light activates an enzyme cascade that stimulates cGMP hydrolysis, leading to channel closure, which blocks calcium influx but not its efflux. Free $[Ca^{2+}]$ in the cell drops within 0.5 s of light stimulation. Recoverin prolongs the photo-response by blocking the phosphorylation of photoexcited rhodoposin by rhodopsin kinase. Isoforms, visinin and neurocalcin, are found in nerve cells; frequenin was characterized in *Drosophila.* Recoverin is N-myristylated. In the crystal structure [21] of non-myristoylated recoverin a single Sm^{3+} ion is bound in place of Ca^{2+} in EF-hand 2. The hydrophobic patches of lobe 1,2 and of lobe 3,4 are exposed and orientated so that their two clefts cannot enfold a target helix as seen in CAM, TNC, ELC, RLC and BM40. NMR studies [22] have shown that the myristoyl side chain is buried between the two lobes in the apo form but is exposed following calcium binding to EF-hands 2 and 3.

Calsenilin (CALS). Calsenilin (CALS) binds both presenilin 1 (PS1) and presenilin 2 (PS2) in a calcium-dependent manner. Most early onset familial Alzheimer cases are caused by mutations in PS1 or PS2. CALS may mediate the role of PS2 in apoptosis and β-amyloid formation [23]. CALS is 251 residues long and contains four EF-hands. No homolog of its first 80 residues has been identified. The sequences of CALS and of DREM are quite similar and all domains cluster together (Fig. 6). They are tentatively placed in different subfamilies because of the apparent differences in function.

DRE-Antagonist Modulator (DREM). The downstream regulatory element (DRE) is known to silence transcription of the prodynorphin and *c-fos* genes. DRE-antagonist modulator (DREM) binds to DRE in the apo form, probably as a homotetramer. "Upon stimulation by Ca^{2+}, DREM's ability to bind to the DRE and its repressor function are prevented" [24]. DREM is 284 residues long and contains four EF-hands. The 110 residues preceding the first EF-hand may be responsible for binding to DNA; however, no homolog is recognized. The first and second EF-hands have unusual distributions of potential Ca^{2+}-binding ligands – X, Y, Z, -X, -Z Asn, Cys, Thr, Asp, Thr and Asp, Asp, Asn, His, Asp. In proteins Ca^{2+} is usually coordinated by O; the S of Cys and the N of His may coordinate water, which in turn binds Ca^{2+}. The third and fourth EF-hands appear to bind calcium.

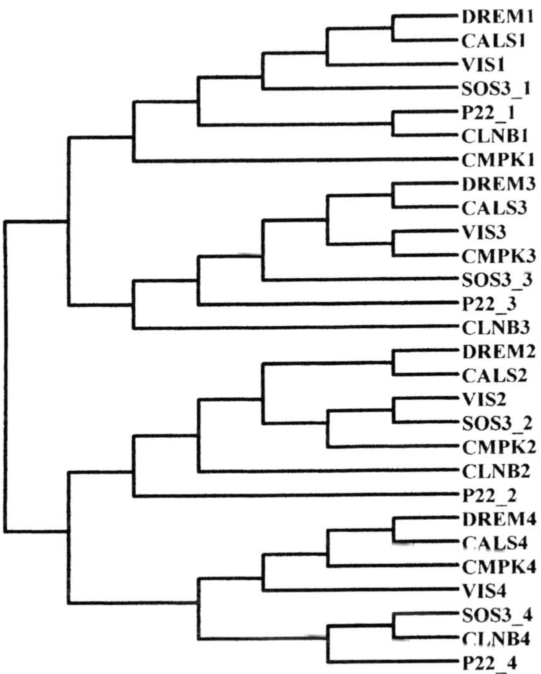

Fig. 6. Dendrogram of CPV. The seven subfamilies – CLNB, *P22*, *VIS*, CALS, DREM, CMPK and SOS3 – show reasonable, but imperfect, congruence. Tentatively we suggest that all seven evolved from a single precursor that contained four EF-hands, a precursor different from the precursor of CTER

Calcium/Calmodulin-Dependent Protein Kinase (CMPK). The calcium/calmodulin-dependent protein kinase (CMPK) of *Lilium* sp. has 520 residues and consists of a serine/threonine protein kinase domain, a calmodulin-binding region, and four EF-hands. CMPK is preferentially expressed in anthers [25]. The mechanism and role of its regulation by calcium is not known.

***sos3* Gene Protein (SOS3).** Mutations in the *sos3* gene make *A. thaliana* hypersensitive to increased Na^+ and to lowered K^+ in the bathing medium [26]. Although SOS3 is congruent to CLNB, P22, VIS, CALS, DREM and CMPK, none of its four EF-hands are inferred to bind calcium.

3.3 Pairings

Reticulocalbin (RTC). Reticulocalbin (RTC), also known as E6BP (E6-binding protein) or ERC-55, appears to bind E6, a transforming protein from human *Papilloma* virus that interacts with tumor suppressor protein p53. RTC has 210 amino acids [27]. It contains six EF-hands, several of which are most similar to the corresponding domains of SCF. RTC is inferred to be in the lumen of the endoplasmic reticulum because it has an N-terminal leader sequence and a glycosylation site. The C-terminal sequence is HDEL; this is similar to the KDEL that serves as a signal to retain proteins within the ER. Four of six EF-hands (1, 4, 5 and 6) bind calcium but 2 and 3 do not [28].

DNA Supercoiling Factor (SCF). DNA supercoiling factor (SCF) is capable of generating negative supercoils into a relaxed DNA in conjunction with eukaryotic DNA topoisomerase II. SCF has 322 residues and contains six EF-hands, five of which are predicted to bind calcium. SCF requires ~10 μM Ca^{2+} for activity. This is inferred to reflect an in vivo response to a calcium signal [29]. Several domains of SCF and RTC are most closely related; however, there is no apparent relation in their functions or homology in the domains at their N-termini.

Calpain (CALP). Calpain (CALP) (EC 3.4.22.17) is an intracellular cysteine protease whose two isoforms require either mM or μM free Ca^{2+} for activity. The large subunit of this heterodimer contains a domain homologous to sulfhydryl protease and five EF-hand domains at the C-terminus. The small subunit contains a Gly-rich domain at its N-terminus and five EF-hands at its C-terminus. In both subunits domains 1 and 2 and 3 and 4 form two lobes. The fifth domains of large and small subunits pair, as usually found in adjacent EF-hands, to join the two in the heterodimer [30].

Sorcin (SORC). Sorcin (SORC) or grancyclin is a homodimer, congruent with CALP, which is inferred to dimerize via its fifth EF-hand. Its concentration is greatly increased in multidrug resistant cells. It has been shown to bind to the ryanodine receptor [31] and to the N-terminal domain of annexin VII (synexin) [32].

S100 (S100). The S100 (S100) family consists of at least 13 distinct isoforms. S100's A1–A10 in humans are all encoded on chromosome 1; whether they are subject to coordinate expression has not been determined. Most have been demonstrated to be dimers. S100a $(S100A1)_2$ regulates giant protein kinase. The NMR solution structure of $(S100B)_2$ free and complexed with the C-terminal peptide (367–388) from protein p53 shows the peptide interacting primarily with α-helix F2 of one monomer and α-helix E1 of the other monomer [33]. S100A9 (p11) forms a heterotetramer with annexin II; the crystal structure of $(S100A9)_2$ complexed peptide 1–13 from annexin II shows a similar orientation of the N-terminal α-helix of the annexin II lying in the 1,2 linker, α-helix F2 groove of one subunit with the N-acetyl group against α-helix E1 of the other subunit [34]. Certain neurons are characterized by their content of S100, as well as their content of parvalbumins. Several S100's function in culture as neurite extension factors [35]; whether S100 functions extracellularly under non-pathological conditions has yet to be established.

Intestinal Calcium-Binding Protein (ICBP). Intestinal calcium-binding protein (ICBP) is strongly congruent with S100 and has a similar structure. It is placed in a different subfamily because it is a monomer; its encoding gene is on the X chromosome, and it is inferred to function in the transcellular transport of calcium [36].

Trichohyalin and Profilaggrin (HYFL). Trichohyalin and profilaggrin (HYFL) have two EF-hands at their N-termini, both of which are congruent with S100 and with ICBP. However, trichohyalin contains seven tandemly repeated domains homologous to involucrin at its C-terminus [37]. Profilaggrin has 10–12 filaggrin domains at its C-terminus [38]. Both associate with keratin intermediate filaments.

Diacylglycerol Kinase (DGK). Diacylglycerol kinase (DGK) (EC 2.7.1.107) has two EF-hands that span residues 114–202. The domain at the N-terminus is not similar to other known protein families. The domain (203–735) at the C-terminus contains an ATP-binding site and cysteine-rich zinc-finger sequences similar to those found in protein kinase C. Isoforms α, β and γ show varying levels of calcium sensitivity [39].

Nucleobindin and DNA-Binding Protein, NEFA (NUBN). Nucleobindin (NUBN) was initially identified in KML_1-7 cells from MRL/l mice, which are prone to develop lupus erythematosis and produce elevated levels of anti-dsDNA and anti-ssDNA antibodies [40]. NUBN has 455 residues. There is a signal sequence at the N-terminus followed by a 200-residue domain of unknown homolgy, two canonical EF-hands, then seven heptad repeats as found in a coiled coil (leucine zipper). The first EF-hands of NUBN and CRGP are most closely related to one another. The nuclear EF-hand acidic (NEFA) gene product is closely related to NUBN; its function is unknown [41].

Calmodulin-Related Gene Product (CRGP). Calmodulin-related gene product (CRGP) is encoded by the gene *T+*, which contains a transposon-like human repeat element, *THE1*, in the 3′ untranslated region of its message [42].

α-Actinin (ACTN). α-Actinin (ACTN) cross links actin filaments non-covalently. It is a homodimer whose subunits assemble in an antiparallel fashion to form a rod-like structure [43]. The difference between nonmuscle and muscle forms is that the cross-linking activity of the former is completely inhibited by μM calcium, while the activity of the muscle form is not calcium sensitive. Both forms have two EF-hand domains at their C-termini.

α-Spectrin (FDRN). α-Spectrin (FDRN) was first described in the erythrocyte in which it reinforces the cell membrane by cross linking ankyrin and protein 4.1. Nonerythroid αII-spectrin and its close homolog α-fodrin form elongated double-stranded filaments. Each strand consists of segments of triple-stranded α-helices that are formed by the strand going forward, back, forward. The two EF-hands at the C-terminus impart a calcium-sensitive [44] interaction with actin and other cytosolic proteins [45]. The first domains of ACTN and FDRN are most closely related; their second EF-hands are also quite similar to one another. Portions of the non-EF-hand regions of ACTN and FDRN appear to be related; the two are placed in distinct subfamilies because of their different interactions with the cytoskeleton.

FAD-Dependent, Mitochondrial Glycerol-3-phosphate Dehydrogenase (GPD). FAD-dependent, mitochondrial glycerol-3-phosphate dehydrogenase (GPD) (EC 1.1.99.5) is located in the outer surface of the inner mitochondrial membrane. The mammalian enzyme is homologous to the yeast and eubacterial FAD-glycerol-3-phosphate dehydrogenases and has two EF-hands at its C-terminus, unlike its yeast and bacterial homologs [46].

Allograft Inflammatory Factor-1 (AIF1). Allograft inflammatory factor-1 (AIF1) is "...a cytokine-inducible, tissue-specific transcript transiently expressed in response to vascular trauma". [47]. It has 143 residues with two EF-hands. Homologs of the first 50 residues have not been identified. The first domains of AIF1 and of GPD are most similar. The arrangement of possible calcium-binding ligands (S-G-G—S-D) of the second EF-hand is unusual.

BM-40 (BM40). BM-40 (BM40), also known as osteonectin or SPARC (secreted protein acidic and rich in cysteines), consists of a short stretch of acidic amino acids, a region homologous to follistatin and an extracelluar calcium (EC) binding module. At least four other extracellular glycoproteins – testican, tsc, QR1 and SC1 – all of unknown function, have homologous follistatin and EC modules. The latter half of the EC module, residues 209–286, consists of a pair of EF-hands (α-helices D, E, F and G of the EC module) as confirmed in the crystal structure [48]. The second EF-hand binds calcium in canonical coordination. The first EF-hand binds calcium in the third (after RLC of molluscs and S100) known variation on the canonical pattern of calcium binding. Since BM40 is extracellular, calcium binding supposedly imparts stability, not information, to it. The cleft between the two EF-hands is occupied by the first α-helix of the EC module, 140–160, in an orientation very similar to that of a target helix bound by calmodulin; this appears to be another example of stabilization, not information transduction.

QR1 (QR1). QR1, named from the cDNA, QR1, cloned from the neuroretina of quail *Coturni coturnix*, has 676 residues. Its C-terminal portion is similar to the extracellular glycoprotein, BM40. QR1 and BM40 are the only examples of (normally) extracellular EF-hand proteins. "Transcription of QR1 takes place only during the late phase of retinal development and is shut off sharply at hatching" [49].

3.4 Self

Onchocerca volvulus **Calcium-Binding Protein (EF12).** A 542-residue protein of unknown function has been isolated from the nematode *Onchocerca volvulus*, as well as its close homolog of 482 residues from *Caenorhabditis elegans* [50]. It has twelve EF-hands, eleven of which conform well to the consensus of Fig. 1 and appear to bind calcium. The fourth EF-hand is highly diverged; however, of all EF-hand domains it most closely resembles EF12 #2. The eleven interdomain spacings are all six residues; this pattern strengthens the interpretation that the 41 (6 + 29 + 6) residues between EF-hands 3 and 5 harbors an aberrant domain. In a dendrogram the six *ODD* domains cluster together and are congruent with the six *EVEN* domains. Further domains 1(2) and 7(8), 3(4) and 9(10), and 5(6) and 11(12) most closely resemble one another, reflecting four cycles of gene duplication and fusion. It is surely significant that the *O. volvulus* and the *C. elegans* proteins align well, with *no* deletions, over the 414 residues (12 * 29 + 11 * 6) encompassing the twelve EF-hands but show many deletions in the N-terminal and C-terminal domains. The six pairs of EF-hands might be arranged helically and undergo a cooperative change in conformation upon binding calcium.

***Lytechinus pictus* SPEC-Resembling Protein (LPS).** *Lytechinus pictus* SPEC-resembling protein (LPS) evolved by a recent gene duplication and fusion from a four-domain precursor, possibly common to the precursor of SPEC. The distribution of introns in the two halves of LPS is similar to one another and to that in SPEC. Further, interdomains 2,3 and 6,7 of LPS are similar to interdomain 2,3 of SPEC [51]. The function of LPS is not known.

Calbindin D_{28k} (CLBN). Calbindin D_{28k} (CLBN) is found in intestinal epithelium and may be involved in the transcellular transport of calcium; however, it is not congruent with ICBP, sometimes referred to as calbindin D_{9k}. Domains 1, 3 and 5 and domains 2, 4 and 6 are more closely related. Interdomains 1,2 and 3,4 and 5,6 are similar and are inferred to be homologous. CLBN appears to have evolved by two duplications and fusions from a two-domain precursor. The six domains, individually synthesized, aggregate in pairwise combination to approximate the spectral properties of the intact CLBN [52].

Gene Product of *eps15* (EP15). The gene product of *eps15* (EP15) encodes a protein of 897 residues. The first domain of 282 residues contains a candidate phosphorylation site as well as six putative EF-hands, two of which are inferred to bind calcium. EP15 is phosphorylated on tyrosine following EGFR activation by EGF in vivo, or directly by EGFR in vitro [53].

***Tetrahymena pyriformis* Calcium-Binding Protein (TCBP).** *Tetrahymena pyriformis* calcium-binding protein (TCBP) may be involved in the calcium-sensitive changes in ciliary beating. The isoforms TCBP-23 and TCBP-25 are 35% identical in amino acid sequence; however, their encoding DNAs are 49% identical [54]. Lobe 1,2 most closely resembles lobe 3,4; this reflects a recent gene duplication and fusion.

P26olf (P26). P26olf (P26) was prepared from the olfactory epithelium of *Rana catesberiana* [55]. P26 has 217 residues. The two putative EF-hands in the first half of P26 are most similar to the two in the second half; as in TCBP this reflects a recent gene duplication. EF-hand 3 is separated from EF-hand 2 by 36 residues that are 60% identical to the C-terminal twenty residues of a rabbit S100. The expectation value of 0.97 indicates that the alignment is not significant for the entire database but this similarity with another EF-hand protein is interesting. None of the EF-hands are inferred to bind calcium in canonical conformation. However, P26 is purified by elution from phenyl Sepharose (following initial binding in 1.0 mM [Ca^{2+}]) with EGTA. It is possible that domains 1 and 3 have an unconventional coordination as found in domains 1 of S100, ICBP and HYFL.

Phospholipase C (PLC). Phospholipase C (PLC) (EC 3.1.4.3) hydrolyzes phosphatidylinositol 4,5-bisphosphate to inositol 1,4,5-trisphosphate and diacylglycerol. It has an absolute requirement for calcium and is activated by guanine nucleotide binding proteins (PLC-β), or by receptor tyrosine kinases (PLC-γ), or by transglutaminase II (PLC-δ). All three vertebrate isoforms of PLC have a plekstrin homolog domain, followed by four EF-hands, a triosephosphate isomerase-like β-barrel catalytic domain, and a C2 domain [56]. Although calcium is absolutely required for activity, it is bound to the catalytic domain. None of the four EF-hands have canonical calcium-binding loops or are Ca^{2+} bound in the crystal structure [57].

CBP1 and CBP2 (CBP). Two similar proteins, CBP1 and CBP2, have been identified in *Dictyostelium discoideum* [58]. Both contain four EF-hands and bind calcium in gel overlay experiments. Expression of CBP1 and CBP2 is increased during development.

3.5 Miscellaneous

***Plasmodium falciparum* Surface Protein (PFS).** *Plasmodium falciparum* surface protein (PFS) has six EF-hand domains and has other sorts of domains at both N- and C-termini [59]. The C-terminal IDEL motif indicates that PFS is found within the endoplasmic reticulum [60].

Calcium-Dependent Protein Kinase (CDPK). Calcium-dependent protein kinase (CDPK) (EC 2.7.1.37) consists of an N-terminal kinase domain (homologous to the b subunit of Ca/CAM-dependent protein kinase II) followed by four EF-hands. A pseudosubstrate sequence and a CAM-binding motif are found at the junction region between the kinase domain and the four EF-hands. A membrane ATPase is one of the substrates of CDPK [61].

***Plasmodium falciparum* Protein Kinase (PFPK).** *Plasmodium falciparum* protein kinase (PFPK) (EC 2.7.1.37) is encoded by the *PfCPK* gene. Residues 56–325 encode a region homologous to CDPK, a protein kinase from plant, with which it shares ~40% sequence identity [62]. The two chimeric kinases probably acquired their respective four EF-hands by different duplication and splicing events.

***Strongylocentrotus purpuratus* Ectodermal Calcium-Binding Protein (SPEC).** *Strongylocentrotus purpuratus* ectodermal calcium-binding protein (SPEC) mRNAs begin to accumulate 20 hours after fertilization in aboral cells [63]. Its function is not known.

p24 Thyroid Protein (TPP). p24 thyroid protein (TPP), or calcyphosine, consists of four EF-hands. Domains 3 and 4 resemble one another most closely. This deviation from the *ODD*, *EVEN* pattern may reflect a recent duplication. The function of TPP is unknown. Another member of the subfamily has been found in lobster and crayfish muscle [64].

1F8 and TB17 (1F8). 1F8 and TB17 (1F8) cDNAs were prepared from highly redundant mRNAs from *Trypanosoma cruzei* [65] and *T. brucei* [66]. 1F8, or calcimedin [67], is located in the flagellum.

Sarcoplasm Calcium-Binding Protein (SARC). Sarcoplasm calcium-binding protein (SARC) binds three Ca^{2+} or Mg^{2+} ions competitively. It is found both as a monomer and as a dimer; it may function as a calcium buffer. Neither the sequences of its four EF-hands nor of the three interdomain linkers closely resemble other EF-hands or linkers. Crystal structures are available of SARC from the sandworm *Nereis diversicolor* [68], and from amphioxus *Branchiostoma lanceolatum* [69]. Although the sequences share only 22% overall identity, the main chain traces of the two are almost superimposable. All four domains have Ca^{2+} bound. The hydrophobic faces of lobes 1,2 and 3,4 contact each another.

Aequorin (AEQ). Aequorin (AEQ) (EC 1.13.12.5 renilla-luciferin 2-monooxygenase) is a calcium-dependent photoprotein that oxidizes luciferin (coelenterazine) and thereby produces the luminescence of the marine coelenterate *Aequoria victoria* [70]. In the anthozoan *Renilla reniformis*, calcium-induced bioluminescence involves two proteins – luciferase- and luciferin-binding protein. LBP has four EF-hands but lacks the luciferase activity of aequorin [71]. AEQ has only a few residues N- and C-terminal to the four EF-hand domains; perhaps its luciferase activity is associated with its aberrant second domain.

Protein Phosphatase (PPTS). Protein phosphatase (PPTS) (EC 3.1.3.16) is encoded by the *rdgC* gene of *D. melanogaster*; it has a phosphoprotein phosphatase domain at its *N*-terminus. Three EF-hands are easily recognized and are indicated in Table 1 as 1, 3 and 4 [72]. Nominal EF-hand 2, occupying 56 residues between domains 1 and 3, may be a far-diverged EF-hand paired with domain 1, another example of guilt by association.

Hra32 (H32). The cDNA clone, Hra32, corresponds to a RNA transcript that accumulates in *Phaseolus vulgaris* (jack bean) during a hypersensitive reaction [73]. The encoded protein, H32, of 161 residues, has four EF-hands, all of which are predicted to bind calcium.

EFH5 (EFH5). EFH5 (EFH5) of trypanosomes is encoded by the *TbEFH5* gene, which is transcribed as a polycistronic messenger. It follows the CAM gene *C* by 111 bp and precedes ubiquitin-EP52/1 by 116 bp [74]. *TbEFH5* is inferred to be under the control of a distant upstream promoter in *T. brucei*, *T. cruzi* and *L. tarentolae* [75].

Calcium Vector Protein (CVP). Calcium vector protein (CVP) of *Amphioxus* is inferred to bind to a target protein. In a model [76] of CVP based on the crystal structures of CAM and TNC, a target α-helix is bound between the two lobes of CVP and the disulfide bond linking α-helices E1 and F2 is accommodated with no steric interference.

PM129 Clone from *Arabidopsis thaliana* (PMAT). PM129 clone from *A. thaliana* (PMAT) cDNA, enriched for plasma membrane associated proteins, contains a single continuous open reading frame [77]. The 37 residues before the first of four EF-hand domains may comprise a distinct domain. Each of the linkers 1,2, 2,3 and 3,4 is eight residues long. There are three Cys residues in the fourth EF-hand and one in the 3,4 linker; a disulfide bond can be formed without distorting the canonical EF-hand. It is strange that domains 1 and 2 of PMAT most closely resemble one another and that domain 4 of PMAT closely resembles domains 1 and 3 of CVP.

LAV1 (LAV). LAV1 (LAV) cDNA encodes a 355-residue protein. Different isoforms are expressed in amoebae and in plasmodia of *Physarum polycephalum* [78]. Domains 2 and 4 of LAV most closely resemble one another. Although LAV appears not to be congruent with any other EF-hand protein, its interdomains 1,2 and 3,4 resemble 1,2 and 3,4 of CTER. Linker 2,3 consists of a sole Leu. This places severe constraints on the spatial relationship between the lobe 1,2 and the C-terminal lobe. The function of LAV is not known and no homolog has been identified for the $\sim$200 residues at its N-terminus.

Calcium-Binding Protein from *Saccharopolyspora erythraea* (CMSE). Calcium-binding protein from *Saccharopolyspora erythraea* (CMSE) is the only known EF-hand protein from eubacteria [79]; none are known from archea. Its function is not known.

MSV (MSV). The entire genome of the pox virus (*Entomopoxvirinae*) of the North American migratory grasshopper, *Melanoplus sanguinipes*, is 236 kbp and contains 267 open reading frames, of which 107 are similar to previously described genes [80]. One gene encodes MSV, 140 residues. MSV has four EF-hands, the first two of which are inferred to bind calcium. It is the only known EF-hand protein from a virus.

Parvalbumin (PARV). Parvalbumin (PARV) consists of three EF-hands, the first of which does not bind calcium and covers the (potential) hydrophobic patch of the lobe formed by the C-terminal two domains. The EF-hands are numbered 2, 3, and 4 because they cluster sequentially with *EVEN*, *ODD*, *EVEN*. The first domain is inferred to have been deleted. EF-hands 3 and 4 of PARV have high affinity for calcium and for magnesium. In the resting muscle cell PARV is in the two-magnesium form. Following a pulse of messenger calcium the bound Mg^{2+} ions dissociate slowly and the excess Ca^{2+} ions are bound by parvalbumin thereby permitting relaxation of skeletal muscle [3]. Whether PARV also performs this relaxing function in other tissues is not known.

Fimbrin (FIMB). Fimbrin (FIMB) is an actin filament bundling protein from chick. It has two EF-hands at its N-terminus and four, 125-residue actin-binding domains, homologous to those of α-actinin, at its C-terminus [81]. The human homologs are I-, L-, and T-plastin whose genes are on chromosomes 3, 13, and X [82]. Even though FIMB bears some similarities to ACTN, their pairs of EF-hands are not closely related.

Ras Guanyl Nucleotide-Releasing Protein (GRP). Ras guanyl nucleotide-releasing protein (GRP) has 795 residues and contains a homolog of a REM domain, a CDC25 domain, two EF-hands, and a diacylglycerol-binding domain [83]. The two EF-hand loops are canonical and are expected to bind calcium. Helices E1 and F2 appear conventional. However, there are only 15 residues for helices F1 and E2. One or both of these helices is truncated and the relationship between the two EF-hands is unclear. Rat fibroblasts expressing the RasGRP encoding gene vector are transformed. Addition of a DAG analog causes sustained activation of Ras-Erk signaling and changes in cell morphology.

***PKD2* Gene Product (PKD).** The *PKD2* gene product (PKD) has 968 residues. Mutations in this gene in humans segregate with polycystic kidney disease [84]. The homolog, PKD1, constitutes a voltage-activated calcium (and sodium) channel in *Caenorhabditis elegans*. PKD has six membrane-spanning helices. Two EF-hands follow the sixth transmembrane helix, 100 residues from the C-terminus.

Ryanodine Receptor Protein (RYR). The ryanodine receptor (RYR) functions as a channel for release of calcium from endoplasmic reticulum of muscle and brain. The isoforms are ~5000 residues long and contain two EF-hands, the second of which is inferred to bind calcium. The channels are open in the presence of $\sim 10^{-6}$ M Ca^{2+} (calcium-induced calcium release) and closed at $\sim 10^{-3}$ M Ca^{2+}. The EF-hands have low calcium affinity and have been suggested to function in closing the channels [85].

Cbl (CBL). CBL binds to phosphorylated tyrosine residues and functions as a negative regulator of many signaling pathways. The crystal structure [86] of CBL (minus residues 1–24) alone and in complex with a phosphotyrosine decapeptide that represents its binding site in ZAP-70 shows a pair of EF-hands wedged between a four-helix bundle (4HB) and a divergent SH2 domain. The coordination of Ca^{2+} is unique, as indicated in Table 1 by "d". Although the residue at -X is Ser, the -X oxygen ligand comes from the carboxylate group of a Glu from the fourth helix of the 4HB. Although calcium is required for phosphopeptide binding, it interacts with the SH2 domain, not the pair of EF-hands. The pair of EF-hands may position the 4HB relative to the SH2 domain.

Calcium- and Integrin-Binding Protein (CIB). Calcium- and integrin-binding protein (CIB) binds specifically to the cytoplasmic domain of integrin α_{IIb} of platelets [87]. It has 191 residues and contains two EF-hands. No homolog has been identified for the first 100 residues.

Calsensin (SENS). Calsensin (SENS) is expressed in some of the peripheral neurons of leech, *Haemopsis* sp., that fasciculate in a single axon tract. SENS has 83 residues and both EF-hands bind calcium [88]. It co-purifies with a 200 kDa protein, not yet characterized

Groovin (GRV). Groovin (GRV) (identical to kakapo) is encoded by *grv* (*kak*) and is expressed in tendon cells of *D. melanogaster*. GRV is essential for muscle-dependent, tendon cell differentiation [89]. GRV contains two EF-hands; both are inferred to bind calcium.

4 Evolution of EF-Hand Proteins

4.1 Ten Subfamilies (CTER) That Resemble Calmodulin

By definition the members of one subfamily resemble one another more closely than they do any member of another subfamily. When comparing subfamilies we initially use one sequence of that subfamily; subsequently conclusions are confirmed by repeating the key comparisons with other representatives, if available, of that subfamily. With minor exceptions the EF-hands 1 (and 2, and 3, and 4) of the ten subfamilies – *C*AM, *T*NC, *E*LC, *R*LC, TPNV, CLAT, SQUD, CDC, CAL and CAST – resemble one another more closely than they do other EF-hands of their own or of other subfamilies. These ten subfamilies are grouped together as "CTER" because, within statistical variation, they are congruent, as illustrated in Fig. 5. CVP, EFH5 and PARV show fair congruence with CTER but are not included. Tentatively we suggest that all ten evolved from a single precursor that contained four EF-hands by a process of gene duplication *without* subsequent gene fusion. That single precursor is inferred to have evolved by two cycles of gene duplication and fusion starting with a single *ur* domain.

Even though the ten are congruent, they show a great divergence in calcium affinity. CLAT and CAST are chimeric, the other eight are not. CAM, TNC, ELC, RLC, TPNV and SQUD all function by activating a target enzyme(s) or structural protein(s). CAM, TNC, ELC and RLC bind an α-helix of the target peptide in the cleft between *ODD* and *EVEN* domains.

4.2 Seven Subfamilies (CPV) That Resemble Calcineurin B

With minor exceptions the EF-hands 1 (and 2, and 3, and 4) of the seven subfamilies – *C*LNB, *P*22, *V*IS, CALS, DREM, CMPK and SOS3 – resemble one another more closely than they do other EF-hands of their own or of other subfamilies. These seven subfamilies are grouped together as "CPV" because, within statistical variation, they are congruent, as illustrated in Fig. 6; however, the degree of congruence is not so strong as for CTER.

The CPV proteins, as was the case for CTER, show a great deal of variation in calcium binding. CALS, DREM and CMPK are heterochimeric; the others are not. CMPK is an enzyme. CLNB binds an α-helix of a target enzyme in the cleft between *ODD* and *EVEN* domains. Despite the similarities, the CTER subfamily EF-hands and the CPV subfamily EF-hands are no more closely related to one another than they are to many EF-hands of other subfamilies. Even so, the postulated *ODD*, *EVEN* precursor common to all members of the CTER group may also be the common precursor of CPV.

4.3 Other General Patterns

Two other patterns, in addition to congruence among CTER and CPV subfamilies, are frequently observed. Seven pairs or triplets of subfamilies most closely resemble the other member(s) of the "Pairings" (triplet) over most of the domains. Often the placement in separate subfamilies, e.g. S100 and ICBP, depends on characteristics other than just amino acid sequence (Table 1).

The "Self" category includes EF12, LPS, CLBN, EP15, TCBP, P26, PLC and CBP. For each subfamily the domain(s) that most closely resembles the EF-hand in question come from the same subfamily. In the most easily interpreted cases EF-hands 1–6 of EF12 (and domains 1–4 of LPS) most closely resemble domains 7–12 of EF12 (5–8 of LPS). EF12 (or LPS) is inferred to have evolved by a recent gene duplication and fusion from a six(four)-domain precursor. This simple pattern of gene duplication and fusion also applies for CLBN, EP15, TCBP and P26. However, for PLC, EF-hands 1 and 2 most closely resemble one another; for CBP, EF-hands 3 and 4 are most similar.

4.4 Introns as Characters for Classification and Evolution

The sequences of genomic DNA are available for 86 different EF-hand sequences in 30 different subfamilies. We have classified these intron sites by phase (between encoding base triplets or after the first and/or after the third base of the triplet) and by position within or between EF-hands into 49 distinct "characters" over the entire lengths of the proteins. These characters can then be scored "+" or "–" for the presence or absence of an intron at that site. The dendrogram based on 29 intron positions (characters) using genomic sequences for 45 CTER proteins for which there are 20 different character states (CAM 15/9, TNC 4/1, ELC 15/2, RLC 7/4, CAL 3/3 and CDC 1/1) is shown in Fig. 7.

The dendrogram based on introns for CAM, TNC, ELC, RLC, CAL and CDC has limited statistical validity because there are only 29 "+/–" characters. Even so, there are several significant features:

1. The intron character states for both human and mouse, skeletal and cardiac TNCs are identical. Over the time scale of vertebrate evolution they have not changed.
2. Most organisms have multiple isoforms of most proteins. The intron characters of two ELCs from *Drosophila*, two RLCs from *Rattus*, and two CAMs from *Halocynthia* are the same. Since the duplication of these encoding genes that gave rise to the isoforms, the introns have not changed.
3. The sharp delineation of the ten subfamilies of CTER based on amino acid sequences as illustrated in Fig. 5 is blurred in the analysis of introns. The TNCs are well resolved from the others. The RLCs of rat smooth muscle and *Drosophila* nonmuscle are very close as are the RLCs of rat skeletal muscle and *Drosophila* muscle; this is consistent with the dendrogram based on amino acid sequence and reflects an early division of the two isoforms.

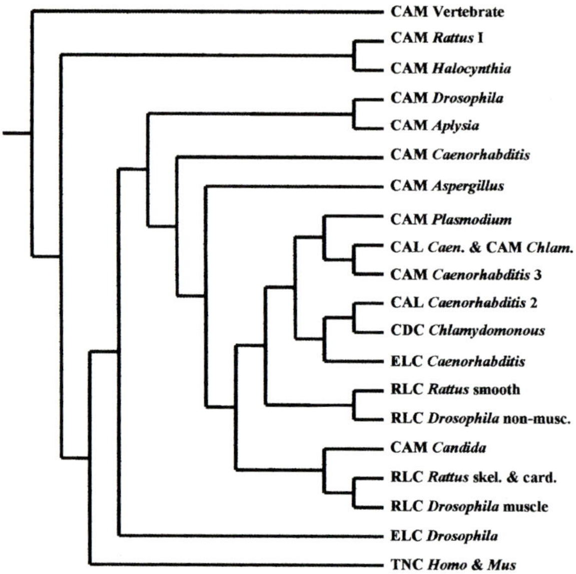

Fig. 7. Dendrogram of CAM, TNC, ELC, CAL and CDC based on intron positions. Sequences of genomic DNA, and hence intron positions, are available for CAM, TNC, ELC, RLC, CAL and CDC. The dendrogram was computed as described in the text. Since the intron distributions are identical for several genomic DNAs, a single taxon may represent several genes in one organism or several different organisms. TNCs from *Homo* and *Mus* both skeletal and cardiac have the same intron character state; two ELCs from *Drosophila*, the same; RLC from *Rattus* skeletal and cardiac muscle; CAM from *Candida* and *Schizosaccharomyces*; CAM from *Chalamydomonas* and CAL *Caenorhabditis*; two CAMs from *Halocynthia*; CAM from *Homo*, *Mus* and *Gallus*

4. The CAL and CDC taxa are more interspersed among the various CAM taxa. For instance, the intron character state of CAL from *Caenorhabditis* and CAM from *Chlamydomonas* are identical. Whether this reflects statistical fluctuation or deeper functional significance is difficult to say; however, the question lends urgency to determining the functions of CAL and of CDC.

5 Perspectives

A chemist might be interested in structure and mechanism with little regard for evolution. In contrast, a biologist might first ask about evolution and only later turn to structure and function. Obviously, a deep understanding of proteins involves a synthesis of both perspectives. Unfortunately, that synthesis is not prescribed by an algorithm. We present several generalizations, offer a few speculations, and admit that we still have a way to go to fulfill the promise of our title.

1. EF-hands almost always occur in adjacent pairs, *ODD* and *EVEN*, and are related by a local, approximate two-fold axis. The two domains fit together well and are

far more stable than an isolated, single EF-hand. The binding of calcium, if bound at all, to the two domains is sometimes cooperative; however, this seems to vary with the protein as well as the experimenter. As exceptions, the fifth domains of CALP (and of SORC) pair with their counterparts to form a dimer. The first in sequence (tabulated as #2) domain of PARV has three residues deleted relative to the canonical EF-hand but does stabilize the 3,4 canonical pair. It is reasonable, in terms of mechanisms of DNA replication and recombination, that any pair of domains from the same family, EF-hands or others, be adjacent. The dendrogram of all 247 subfamily domains shows a strong clustering of odd-numbered domains in one half and even in the other half. Whether this reflects the influence of the linker between the *ODD* and *EVEN* EF-hands, in contrast to the absence of covalent coupling from *EVEN* to *ODD*, or whether it reflects the trace of the original gene duplication event and subsequent random divergence, remains to be resolved.

2. About a third of EF-hands are inferred or observed not to bind calcium. Further there are at least four non-canonical arrangements of calcium-binding ligands. Surely the ability to bind or not to bind calcium was lost and gained multiple times during evolution, an example of homoplasy at the molecular level. We have no idea whether the *ur*-EF-hand was able to bind calcium.
3. Of the 66 subfamilies, 31 are heterochimeric. Domains, other than EF-hands, are spliced N-ward or C-ward to the adjacent EF-hands. Eight have other domains both N-ward and C-ward to the EF-hands and P26 has a different domain between EF-hand pair 1,2 and pair 3,4. The functions of many of these heterochimeras are not known; however, at least seven of them – CALP, GPD, DGK, PLC, CDPK, PPTS and PFPK (*Plasmodium*), are cytosolic enzymes whose functions are known or inferred to be modulated by calcium. These enzymes may "want" to have their own transducers on board so their responses can differ from those proteins that interact with a general transducer like CAM. Perhaps if we returned in 10^8 years, we would find a CLNB spliced to a calciuneurin A.
4. Several of the 39 (31 + 8) non-EF-hand regions consist of multiple domains. Few, if any, of these 39 domains are related to one another. The introns of EF-hand proteins occur as frequently within the EF-hand domains as before, between, or after them. The gene splicings that generated the multiple EF-hands and the 31 heterochimeras were indiscriminately promiscuous; we discern no preference or pattern.
5. There are at least 1000 distinct protein domains, probably less than 10,000 in the biosphere. All members of a domain family are assumed to have evolved from a common origin, that is, to be homologous. There are several superfamilies in which the constituent families may be homologs or in which the resemblance of families reflect divergent evolution. We infer, but certainly cannot prove, that all 66 subfamilies are homologs. However, the interpretation of homology is especially tenuous and of limited statistical significance when dealing with only 30 residues in a helix, loop, helix structure. The database contains lots of helices related by $\sim 90°$; hence, our reliance on additional characteristics like function calcium binding, domain pairing, and congruence, all of which have known exceptions.

6. The existence of domains that resemble EF-hands raises two questions that apply in varying degrees to all domain families. At the limit can one be certain that every member of a family more closely resembles every other member of its own homolog family than any member of any other family? **No**! However, the family assignments are usually correct and the attendant classifications, however imperfect, are valuable both in tracing evolution and suggesting functional and structural characteristics.
7. Of greater interest is the question of the origin of proteins, as opposed to the origin of life. Certainly the first cell did not appear replete with a 1000 distinct protein domains. Yet the 10^3 or 10^4 contemporary domain families reflect the limit of our vision. How did the first EF-hand arise? And why, apparently, in eukaryotes? We can hardly speculate as to whether it branched off a fragment of a globin fold or from a zinc finger. Perhaps it arose de novo in the sense that some bit of non-coding (junk) DNA that potentially encoded a couple of helices acquired an initiation sequence. Few people are bold, or foolish, enough to speculate as to how the 1000 domains arose. But this is a fascinating and legitimate question. Hopefully, detailed analyses of individual protein families will provide foundations and boundaries for such theories.

References

1. Kretsinger RH (1972) Nature New Biology 240:85
2. Pearson WR (1990) Rapid and sensitive sequence comparison with FASTP and FASTA. In: Doolittle RF (ed) Pp.63 in Methods in Enzymology. Academic Press, San Diego New York, p 736
3. Gilles J-M, Thomason DB, LeFevre J, Kretsinger RH (l982) J Muscle Res Cell Motil 3:377
4. Einspahr H, Bugg CE (1984) Metal Ions Biol Systems 17:52
5. Yap KL, Ames JB, Swindells MB, Ikura M (1999) Proteins 37:1
6. Rhoads A, Friedberg F (1997) FASEB J 11:331
7. Finn BE, Forsén S (1995) Structure 3:7
8. Vassylev DG, Takeda S, Wakatsuki S, Maeda K, Maeda Y (1998) Proc Natl Acad Sci USA 95: 4847
9. Fromherz S, Szent-Gyorgyi AG (1995) Proc Natl Acad Sci USA 92:7652
10. Houdusse A, Cohen C (1996) Structure 4:21
11. Garone L, Theibert JL, Miegel A, Maeda Y, Murphy C, Collins JH (1991) Arch Biochem Biophys 291:89
12. Ling V, Zielinski RE (1993) Plant Mol Biol 22:207
13. Bearer EL, DeGiorgis JA, Jaffer H, Medeiros NA, Reese TS (1996) Proc Natl Acad Sci USA 93:6064
14. Biggins S, Ros MD (1994) J Cell Biol 125:843
15. Salvato M, Sulston J, Albertson D, Brenner S (1986) J Mol Biol 190:281
16. Gellatly KS, Lefebvre DD (1993) Plant Physiol 101:1405
17. Kissinger CR, Parge HE, Knighton DR, Lewis CT, Pelletier LA, Tempczyk A, Kalish VJ, Tucker KD, Showalter RE, Moomaw EW, Gastinel LN, Hakuba N, Chen X, Maldonado F, Barker JE, Bacquet R, Villafranca JE (1995) Nature 378:641
18. Griffith JP, Kim JL, Kim EE, Sintchak MD, Thomson JA, Fitzgibbon MJ, Fleming MA, Caron PR, Hsiao K, Navia MA (1995) Cell 82:507
19. Barroso MB, Bernd KK, DeWitt ND, Chang A, Mills K, Stzul ES (1996) J Biol Chem 271:10183
20. Lin X, Barber DL (1996) Proc Natl Acad Sci USA 93:12631
21. Flaherty KM, Zozulys S, Stryer L, McKay DB (1993) Cell 75:709
22. Ames JB, Porumb T, Tanaka T, Ikura M, Stryer L (1995) J Biol Chem 270:4526
23. Baxbaum JD, Choi E-K, Luo Y, Lilliehook C, Crowley AC, Merriam OE, Wasco W (1998) Nature Med 4:1177

24. Carrión AM, Link WA, Ledo F, Mellstrom B, Naranjo JR (1999) Nature 398:80
25. Patil S, Takezawa D, Poovaiah BW (1995) Proc Natl Acad Sci USA 92:4897
26. Liu J, Zhu J-K (1998) Science 280:1943
27. Chen JJ, Reid CE, Band V, Androphy EJ (1995) Science 269:529
28. Tachikui H, Navet AF, Ozawa M (1997) J Biochem (Tokyo) 121:145
29. Ohta T, Kobyashi M, Hirose S (1995) J Biol Chem 270:15571
30. Hohenester E, Mauer P, Hohenadl D, Timpl R, Jansonius JN, Engel J (1996) Nature Struct Biol 3:67
31. Lokuta AJ, Meyers MB, Sander PR, Fishman GI, Valdivia HH (1997) J Biol Chem 272:25333
32. Brownawell AM, Creutz CE (1997) J Biol Chem 272:22182
33. Rustandi RR, Drohat AC, Weber DJ (1998) Biochemistry 37:1951
34. Rety S, Sopkova J, Renouard M, Osterloh D, Gerke V, Tabaries S, Russo-Marie F, Bentley A (1999) Nature Struct Biol 6:89
35. Heizmann CW, Cox JA (1996) Bio Metals 11:383
36. Wooding FBP, Morgan G, Jones GV, Care AD (1996) Cell Tissue Res 285:477
37. Lee S-C, Kim I-G, Marekov LN, O'Keefe EJ, Parry DAD, Steinert PM (1993) J Biol Chem 268:12164
38. Markova NG, Marekov LN, Chipev CC, Gan S-Q, Idler WW, Steinert PM (1993) Mol Cell Biol 13:613
39. Yamada K, Sakane F, Matsushima N, Kanoh H (1997) Biochem J 321:59
40. Miura K, Titani K, Kurosawa Y, Kanai Y (1992) Biochem Biophys Res Commun 187:375
41. Den Z, Pelias MZ, Ditta GM, Keats BJ, Deininger PL, Batzer MA (1998) Biochim Biophys Acta 1407:84
42. Deka N, Wong E, Matera AG, Kraft R, Leinwand LA, Schmid CW (1988) Gene 71:123
43. Taylor KA, Taylor DW (1993) J Mol Biol 230:196
44. Lundberg S, Buevich AV, Sethson I, Edlund U, Backman L (1997) Biochemistry 23:7199
45. Djinovic-Carugo K, Young P, Gautel M, Saraste M (1999) Cell 98:537
46. Brown LJ, MacDonald MJ, Lehn DA, Moran SM (1994) J Biol Chem 269:14363
47. Autieri MV (1996) Biochem Biophys Res Commun 228:29
48. Hohenester E, Mauer P, Hohenadl D, Timpl R, Jansonius JN, Engel J (1996) Nature Struct Biol 3:67
49. Guermah M, Crisanti P, Laugier D, Dezelle P, Bidou L, Pessac B, Calothy G (1991) Proc Natl Acad Sci USA 88:4503
50. Lizotte-Waniewski M, Tawe W, Guiliano DB, Lu W, Liu J, Williams SA, Lustigman S (2000) Infec Immun, in press
51. Xiang M, Ge T, Tomlinson CR, Klein WH (1991) J Biol Chem 266:10524
52. Linse S, Thulin E, Gifford LK, Radzewsky D, Hagan J, Wilk RR, Åkerfeldt KS (1997) Prot Sci 6:2385
53. Fazioli F, Minichiello L, Matoskova B, Wong WT, Di Fiore PP (1993) Mol Cell Biol 13:5814
54. Takemasa T, Takagi T, Kobayashi T, Konishi K, Watanabe Y (1990) J Biol Chem 265:2514
55. Miwa N, Kobayashi M, Takamaztsu K, Kawamura S (1998) Biochem Biophys Res Commun 25:860
56. Grobler JA, Hurley JH (1998) Biochemistry 37:5020
57. Essen L-O, Perisic O, Cheung R, Katan M, Williams RL (1996) Nature 380:595
58. André B, Noegel AA, Schleicher M (1996) FEBS Lett 382:198
59. Templeton TJ, Fujioka H, Aikawa M, Parker KC (1997) Mol Biochem Parasitol 90:359
60. La Greca N, Hibbs AR, Riffkin C, Roley M, Tilley L (1997) Mol Biochem Parasitol 89:283
61. Saijo Y, Hata S, Sheen J, Izui K (1997) Biochim Biophys Acta 1350:109
62. Zhao Y, Kappes B, Franklin RM (1993) J Biol Chem 268:4347
63. Hardin PE, Angerer LM, Hardin SH, Angerer RC, Klein WH (1988) J Mol Biol 202:417
64. Sauter A, Staudenmann W, Hughes GJ, Heizmann CW (1995) Eur J Biochem 227:97
65. Engman DM, Krause K-H, Blumin JH, Kim KS, Kirchhoff LV, Donelson JE (1989) J Biol Chem 264:18627
66. Lee MG-S, Chen J, Ho AWM, D'Alesandro PA, Van der Ploeg LHT (1990) Nucleic Acids Res 18:4252
67. Wu Y, Haghighat NG, Ruben L (1992) Biochem J 287:187
68. Cook WJ, Falk SE, Babu YS, Cox JA, Vijay Kumar S (1991) J Biol Chem 266:652

69. Cook WJ, Jeffrey LC, Cox JA, Vijay Kumar S (1993) J Mol Biol 229:461
70. Prasher DC, McCann RO, Longiaru M, Cormier MJ (1987) Biochemistry 26:1326
71. Kumar S, Harrylock M, Walsh KA, Cormier MJ, Charbonneau H (1990) FEBS Lett 268:287
72. Steele FR, Washburn T, Rieger R, O'Tousa JE (1992) Cell 69:669
73. Jakobek JL, Smith-Becker JA, Lindgren PB (1999) Mol Plant Microbe Interact 12:712
74. Ajioka J, Swindle J (1993) Mol Biochem Parasitol 57:127
75. Wong S, Morales TH, Neigel JE, Campbell DA (1993) Mol Cell Biol 13:207
76. Cox JA, Alard P, Schaad O (1990) Prot Engineering 4:23
77. Bartling D, Bülter H, Weiler EW (1993) Plant Physiol 102:1059
78. Laroche A, Lemieux G, Pallotta D (1989) Nucleic Acids Res 17:10502
79. Swan DG, Cortes J, Hale RS, Leadlay PF (1989) J Bacteriol 171:5614
80. Afonso CL, Tulman ER, Lu Z, Oma E, Katish GF, Rock DL (1999) J Virol 73:533
81. de Arruda MV, Watson S, Lin C-S, Leavitt J, Matsudaira P (1990) J Cell Biol 111:1069
82. Lin CS, Shen W, Chen ZP, Tu YH, Matsudaira P (1994) Mol Cell Biol 14:2457
83. Ebinu JO, Bottorff DA, Chan EYW, Stang SL, Dunn RJ, Stone JC (1998) Science 280:1082
84. Mochizui T et al. (1996) Science 272:1339
85. Xiong H, Feng X, Gao L, Xu L, Pasek DA, Seok JH, Meissner G (1998) Biochemistry 37:4804
86. Meng W, Sawasdikosol S, Burakoff S, Eck MJ (1999) Nature 398:84
87. Naik UP, Patel PM, Parise LV (1997) J Biol Chem 272:4651
88. Briggs KK, Silvers AJ, Johansen KM, Johansen J (1995) J Cell Biol 129:1355
89. Strumpf D, Volk T (1998) J Cell Biol 143:1259

Calmodulin Target Recognition: Common Mechanism and Structural Diversity

Tao Yuan, Kyoko L. Yap, Mitsuhiko Ikura*

Division of Molecular and Structural Biology, Ontario Cancer Institute and Department of Medical Biophysics, University of Toronto, 610 University Avenue, Toronto Ontario M5G 2M9, Canada (*E-mail: mikura@oci.utoronto.ca*)

Calmodulin is an intracellular calcium receptor which regulates the biological activities of a wide variety of proteins including protein kinases, phosphatases, and ion transporters. During the past decade, three-dimensional structures of calmodulin in forms of various target complexes have been elucidated. This review presents the latest advances in the field of calmodulin and its target proteins, with particular emphasis on the structural basis underlying multi-target recognition processes conducted by this Ca^{2+}-dependent cellular activator.

Keywords. Signal transduction, Calcium signaling, Three-dimensional structure, NMR, X ray, Molecular recognition

List of Abbreviations

CaM	calmodulin
CaMKI	calmodulin-dependent protein kinase I
CaMKII	calmodulin-dependent protein kinase II
CaMKK	calmodulin-dependent protein kinase kinase
CD	circular dichroism
MARCKS	myristoylated alanine-rich C kinase substrate
MLCK	myosin light chain kinase
NMR	nuclear magnetic resonance
nOe	nuclear Overhauser effect
NOS	nitric oxide synthase
PGD	plant glutamate decarboxylase
SAXS	small angle X-ray scattering
TFP	trifluoperazine
W-7	*N*-(6-aminohexyl)-5-chloro-1-naphthalene sulfonamide

1 Introduction

The calcium (Ca^{2+}) concentration inside a eukaryotic cell is tightly regulated. While the Ca^{2+} concentration in extracellular space is around 10^{-3} mol/l, the intracellular Ca^{2+} concentration at the resting state is about 10^{-7} mol/l. Upon activation by various external stimuli, Ca^{2+} ions influx into the cytoplasm from an extracellular Ca^{2+} pool or intracellular Ca^{2+} stores such as the endoplasmic reticulum or sarcoplasmic reticulum. As a consequence, the intracellular Ca^{2+} concentration raises to around the 10^{-6} to 10^{-5} mol/l range. The increase of Ca^{2+} concentration in the cytoplasm results in Ca^{2+} binding and activation of a family of Ca^{2+}-binding proteins, known as the "EF-hand" (helix-loop-helix) protein family [1]. Among the ~200 currently known members, calmodulin (CaM) is unique in that it can activate numerous target proteins upon Ca^{2+} binding. CaM is a small (molecular weight around 16,700), acidic protein which contains 148 residues (Fig. 1). It is one of the most conserved among eukaryotic proteins [2]. Surprisingly, this small protein can bind and activate more than 100 different target proteins, including myosin light chain kinases (MLCK), CaM-dependent protein kinases (CaMK), phosphatase 2B (calcineurin), nitric oxide synthase (NOS), plasma membrane Ca^{2+} pump, and various ion channels [3–5]. By activating these enzymes and proteins, CaM regulates a variety of physiological processes including cell cycle control, muscle contraction, transcription, cyclic nucleotide metabolism, nitric oxide production and ion movement [6]. There are many reviews describing CaM's physiological functions as well as its molecular and structural properties [3–11]. In recent years, a number of structural studies have been reported for CaM in complex with target peptides and inhibitors, enriching our understanding of the mechanisms underlying CaM target recognition. Here we focus on the structural aspect of CaM and its target and inhibitor complexes.

2 Structure of Ca^{2+}-Loaded Calmodulin

The first X-ray structure of Ca^{2+}-loaded CaM (Ca^{2+}-CaM) was reported more than a decade ago [12, 13]. Since then, many X-ray structures have been determined at higher resolution for Ca^{2+}-CaM from various species, such as mammals [14, 15], *Drosophila melanogaster* [16] and *Paramecium tetraurelia* [17, 18]. These Ca^{2+}-CaM crystal structures are very similar to each other, possessing an elongated dumbbell-shape, with an overall size of about 60 × 30 × 30 Å (Fig. 2). The two

EF-hand		1st helix	loop	2nd helix	
			1 3 5 7 9 12 X Y Z -Y -X -Z		
I	ADQLTEEQIA	EFKEAFSLF	DKDGDGTITTKE	LGTV**M**RSL	39
II	GQNPTEA	ELQD**M**INEV	DADGNGTIDFPE	FLT**MM**ARK	75
III	**M**KDTDSEE	EIREAFRVF	DKDGNGYISAAE	LRHV**M**TNL	112
IV	GEKLTDE	EVDE**M**IREA	DIDGDGQVNYEE	FVQ**MM**TAK	148

Fig. 1. The amino acid sequence of mammalian CaM. The sequences of four "helix-loop-helix" EF-hand Ca^{2+}-binding motifs are aligned. The coordinating residues for the Ca^{2+} ion within the EF-hand loop are indicated *at the top of the sequence*. The unstructured part of the central linker region of Ca^{2+}-CaM in solution is *underlined* (D78-S81). Met residues which are important for target binding are in *bold*

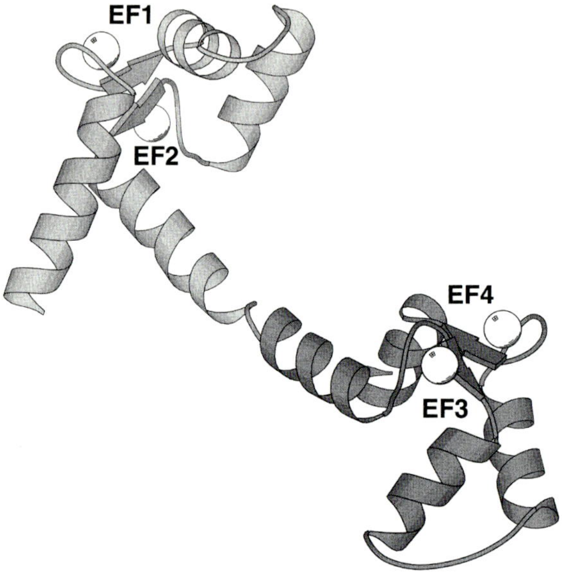

Fig. 2. The crystal structure of Ca^{2+}-loaded CaM [12]

domains of CaM are connected by a solvent-exposed, 28-residue, seven-turn α-helix. There is no direct contact between the two domains of Ca^{2+}-CaM in the X-ray structures. The homologous domains comprise residues 1–77 for the amino-terminal (N-) domain and 78–148 for the carboxy-terminal (C-) domain, as determined from limited trypsin digestion of Ca^{2+}-CaM [19, 20]. On the surface of each domain is a hydrophobic cavity composed of 14 hydrophobic residues (Met, Phe, Ile, Val, Leu) (Fig. 3) with some negatively charged residues (Glu and Asp) at the rim of the cavity [14]. The hydrophobic cavity of each domain is approximately 10.0 × 12.5 Å with a depth of about 9.5 Å. These two exposed hydrophobic cavities are responsible for CaM's interaction with its target proteins [21, 22], which will be discussed below.

The basic structural block of CaM is the EF-hand Ca^{2+}-binding motif consisting of 29 residues: 10 residues for the first α-helix, 12 residues for the calcium-binding loop and 10 residues for the second α-helix, with 3 residues shared by the loop and the second α-helix (Fig. 1). Each domain contains four α-helices forming a pair of Ca^{2+}-binding EF-hand motifs, which are connected by a small β-sheet between the two Ca^{2+}-binding loops. The eight helices of the four EF-hands in CaM are helix A (residues 5–19), helix B (29–38), helix C (45–55), helix D and E (65–92), helix F (102–111), helix G (118–128) and helix H (138–147). The two β-sheets constitute residues 26–28 and 62–64 in the N-terminal domain, and 99–101 and 135–137 in the C-terminal domain. Hydrogen bonds are observed between the backbone nitrogen and oxygen atoms of the middle residues (8th position in the Ca^{2+}-binding loop) of the small β-sheet in each domain. In each of the four calcium-binding loops (residues 20–31, 56–67, 93–104 and 129–140; see Fig. 1), Ca^{2+} is coordinated by seven oxygen atoms of a distorted pentagonal bipyramid with three side-chain oxygen atoms from the first, third and fifth positions of the loop, a backbone carbonyl

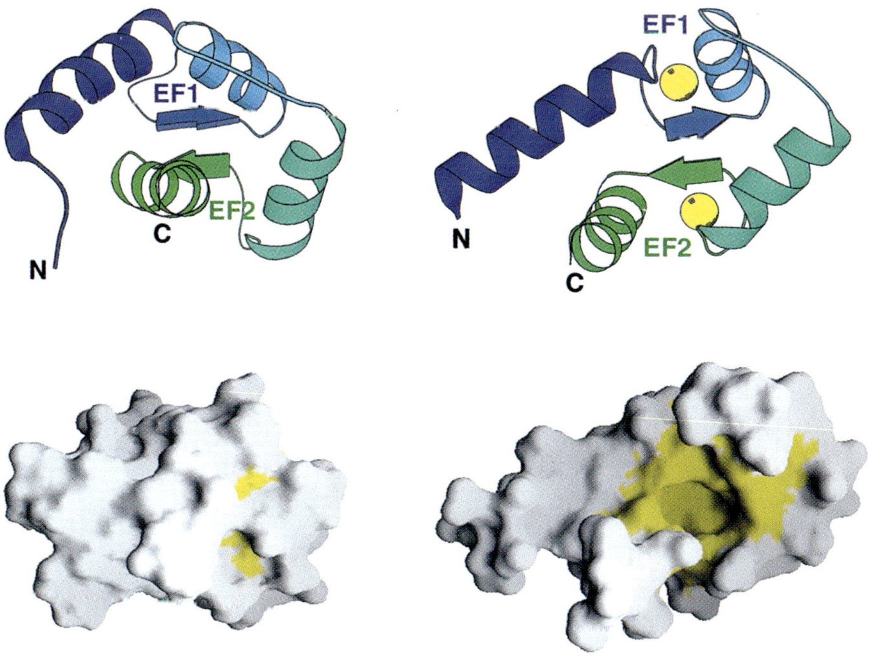

Fig. 3. Ca^{2+}-induced conformational change of CaM. The N-terminal domain containing EF1 and EF2 is shown. In the *upper panels*, ribbon representation of the polypeptide backbone is shown for apo-CaM [31] (*left*) and Ca^{2+}-loaded CaM [12] (*right*). In the *lower panels*, surface representation is shown for apo-CaM (*left*) and Ca^{2+}-loaded CaM (*right*), with hydrophobic residues *highlighted in yellow*

oxygen from the seventh position, a water molecule and two side-chain oxygen atoms from an invariant Glu at the twelfth position (Fig. 1). In addition, there are hydrogen bonds formed between residues at positions 1 and 4, 3 and 12, and 9 and 12. The calcium-oxygen distance is typically around 2.3–2.4 Å.

The long, solvent-exposed central helix observed in all X-ray structures of Ca^{2+}-CaM has been of great interest to researchers in the field. The temperature factor of this region is nearly twice as high as the average for the other parts of the molecule (excluding the N- and C-termini), with the middle residues having the highest temperature factors. Distortions or kinks are also observed in different structures in this region [14, 16, 17]. Furthermore, significant deviation of dihedral angles from normal α-helix values are observed for several residues in this central α-helix. The above observations suggested that the central α-helix possesses intrinsic flexibility. In addition, earlier limited trypsin digestion studies showed that CaM can be cleaved into two halves, suggesting an existence of a trypsin-accessible uncoiled region in the central α-helix [19, 20]. Using small angle X-ray scattering (SAXS) techniques, Heidorn and Trewhella [23] showed that the two domains of Ca^{2+}-CaM are closer to each other than the distance found in X-ray structures, suggesting that the domain linker is not completely α-helical in solution. The flexible nature of the central α-helix was also confirmed by hydrogen-deuterium exchange studies, nuclear

Overhauser effects (nOe) and chemical shift analysis by nuclear magnetic resonance (NMR) spectroscopy [24, 25]. Finally, Barbato et al. [26] demonstrated conclusively by NMR relaxation studies that residues 78–81 are highly mobile in solution and are not in an α-helical conformation. The uncoiled central linker region in CaM causes its two domains to tumble independently in solution. The flexible central linker region is one of the key structural features that enable CaM to bind and activate a range of target proteins (see below). One possible reason for the observation of a long α-helix in this region in the X-ray structures may be due to the organic component (2-methyl 2,4-pentanediol) of the crystallization buffer, since such an alcoholic solvent may induce or stabilize α-helices in proteins [27].

3
Structure of Apo-Calmodulin

In 1986, Herzberg et al. [28] proposed a model for CaM's conformational changes based on the structure of troponin C, in which the C-terminal domain was occupied with Ca^{2+} but the N-terminal domain was free of metal ions. The details of this conformational transition was, however, uncertain until the structure of apo-CaM was solved by NMR spectroscopy in 1995 [29–31]. As in the case of Ca^{2+}-CaM, the central linker region of apo-CaM is not well defined because of its intrinsic flexibility [30, 31]. On the other hand, each domain of apo-CaM is well defined, with the same secondary structural elements observed in Ca^{2+}-CaM. Two Ca^{2+}-binding EF-hand motifs are joined together by a small β-sheet between the two Ca^{2+}-binding loops, in a similar manner to that observed in Ca^{2+}-CaM. Despite the conservation of secondary structure between the Ca^{2+}-free and -bound states, each domain differs significantly in helix packing between the two states. In apo-CaM, both domains assume the "closed conformation" in which the two helices of each EF-hand are positioned nearly anti-parallel to one another (interhelical angle around 128–137°).

Upon binding Ca^{2+}, the two helices become more perpendicular (around 86–101°), each of the domains exhibiting the "open conformation" [31] (Fig. 3). Alternatively, the Ca^{2+}-induced change may be regarded as the movement undergone by helices B and C, while helices A and D remain fixed, as has been observed in both domains of CaM [29–31]. In addition to this closed-to-open conformational transition, a variety of helix movement has been observed in other Ca^{2+}-regulatory proteins, such as spectrin, recoverin, troponin C and S100 proteins [32–36]. The conformational changes of these EF-hand proteins, which vary in both magnitude and quality, have recently been analyzed by a technique termed the Vector Geometry Mapping (VGM) method [37]. In contrast, calbindin, a Ca^{2+}-buffering protein, experiences little tertiary structure arrangement upon Ca^{2+}-binding [38].

On its surface, apo-CaM has few hydrophobic residues and a large number of negatively charged residues (Fig. 3). This surface may be responsible for Ca^{2+}-independent binding of CaM to other target proteins, albeit with much lower affinity [31]. Interactions between apo-CaM and its targets are mainly electrostatic in nature, since they are sensitive to the salt concentration in solution [39].

Tjandra et al. [40] have shown that, for the C-terminal domain of apo-CaM, two sets of conformers are detected in the time scale of several hundred microseconds when in solution. The C-terminal domain of apo-CaM seems to be more flexible in backbone mobility than the N-terminal domain [30, 31, 40, 41]. Using a molecular modeling approach, Swindells and Ikura [42] proposed that the C-terminal domain of apo-CaM assumes a semi-open conformation, which was supported recently by experimental methods [43]. The structural and dynamic features of apo-CaM may account for its C-terminal domain being a primary binding site for its target proteins in the absence of calcium [39, 44].

4
Ca^{2+}-Calmodulin-Target Complexes

The CaM binding motif of target proteins typically comprises a stretch of 20–25 amino acids which are mainly hydrophobic and positively charged. The CaM binding motif often has the propensity to form an α-helix upon binding to CaM. Biophysical and structural properties of known CaM binding motifs have been summarized by Rhoads and Friedberg [5] and in the Calmodulin Target Database (http://calcium.oci.utoronto.ca). Chemically synthesized peptides derived from the target protein's CaM binding sites have been used extensively to study CaM-target interactions, since CaM binding sites are localized and the affinity of CaM for these peptides is comparable to that for the intact proteins.

4.1
Ca^{2+}-Calmodulin-MLCK Complex

MLCK is a serine/threonine protein kinase involved in muscle contraction (for a recent review, see [45]). The C-terminus of the catalytic domain extends to an autoinhibitory domain which overlaps with the CaM binding domain. Binding of Ca^{2+}-CaM to the CaM binding domain induces the displacement of the autoinhibitory domain from the active site within the catalytic core, resulting in activation of the target enzyme. In this manner, MLCK is able to bind and subsequently phosphorylate its substrate, myosin light chain.

In 1992, the first three-dimensional (3-D) structure of Ca^{2+}-CaM complexed with a 26-residue target peptide derived from skeletal muscle myosin light chain kinase (skMLCK) was determined by multidimensional NMR spectroscopy [46]. The CaM-MLCK peptide complex assumes a globular shape with approximate ellipsoid dimension of $47 \times 32 \times 30$ Å (Fig. 5A), in contrast to the dumbbell-shaped X-ray structure of Ca^{2+}-CaM. The two domains come together to engulf the MLCK peptide with the aid of a long flexible loop between the domains (residue 74 to 82). The two domains of CaM in the complex possess approximate two-fold symmetry. The MLCK peptide experiences a transition from a random coil to an α-helix upon binding to Ca^{2+}-CaM. In the complex, more than 80% of the surface area of the MLCK peptide is buried (1477 Å^2) and the majority of the interface provided by Ca^{2+}-CaM (1848 Å^2) is hydrophobic in nature [46].

The hydrophobic surface of each domain in CaM has an unusually high content of methionine residues (M36, M51, M71 and M72 in the N-terminal domain; M109, M124, M144 and M145 in the C-terminal domain). All these Met residues interact with hydrophobic residues in the MLCK peptide. The highly flexible nature and polarizable sulfur atom of the Met side chain have been thought to be crucial for CaM to bind various targets [47]. In addition to the hydrophobic interactions, exquisite electrostatic interactions are found between Asp and Glu residues in Ca^{2+}-CaM and Arg and Lys residues in the skMLCK peptide. For example, R344 of the peptide is close to E127 of CaM, and K346 and K347 are near E11 and E14. Finally, the MLCK peptide uses two bulky hydrophobic residues: W345 binds deeply in the hydrophobic cavity of the C-terminal domain of Ca^{2+}-CaM, while F358 interacts with the hydrophobic cavity of the N-terminal domain (Fig. 4, 6A). These interactions result in the antiparallel orientation of the peptide with respect to Ca^{2+}-CaM: the N-terminal half of the peptide predominantly interacts with the C-terminal domain of CaM and the C-terminal half of the peptide interacts with the N-terminal domain of CaM. This binding mode of using two bulky hydrophobic residues separated by 12 residues represents one class CaM target recognition (Fig. 4).

A 2.4-Å X-ray structure was subsequently reported for a complex between Ca^{2+}-CaM and its binding domain from smooth muscle MLCK (smMLCK) [48]. This structure is similar to the NMR structure of the Ca^{2+}-CaM-skMLCK peptide complex. A minor difference was found in the flexible loop in the central linker region: the former structure includes residues 73–77, whereas the latter involves residues 74–82. Meador et al. [48] also found two bulky hydrophobic residues, W800 and L813 (corresponding to W345 and F358 in skMLCK), anchoring the peptide to the hydrophobic cavity of C- and N-terminal domains of CaM, respectively. This finding again emphasizes the importance of two large hydrophobic residues in CaM target recognition (Fig. 4).

4.2 Ca^{2+}-Calmodulin-CaMKII Complex

In 1993, a new class of CaM-peptide interaction was elucidated by the X-ray structure of Ca^{2+}-CaM complexed with a peptide derived from CaM-dependent protein kinase II (CaMKII) [49]. This multi-functional protein kinase plays an important role in transcriptional regulation (for a recent review, see [50]). In the crystal structure, the

```
skMLCK                     342- K R R W K K N F I A V S A A N R F K K I S S S G A L -367
smMLCK                   796- A R R K W Q K T G H A V R A I G R L S S -815
Ca2+-Pump C24W            1104- Q I L W F R G L N R I Q T Q I R V V N A F R S S -1127
          C20W    1100- L R R G Q I L W F R G L N R I Q T Q I K   -1119

CaMKI                    299- A K S K W K Q A F N A T A V V R H M R K L Q -320
Titin kinase             282- Y H T L I K K D L N M V V S A A R I S C G G -303

CaMKII         290- L K K F N A R R K L K G A I L T T M L A T R N F S -314

rCaMKKα            438- V K L I P S W T T V I L V K S M L R K R S F G N P F -463
cCaMKK             331- V R V I P R L D T L I L V K A M G H R K R F G N P F R -357
```

Fig. 4. The amino acid sequences of CaM binding sites in target proteins. Critical anchoring residues in each sequence are *boxed*

a b c

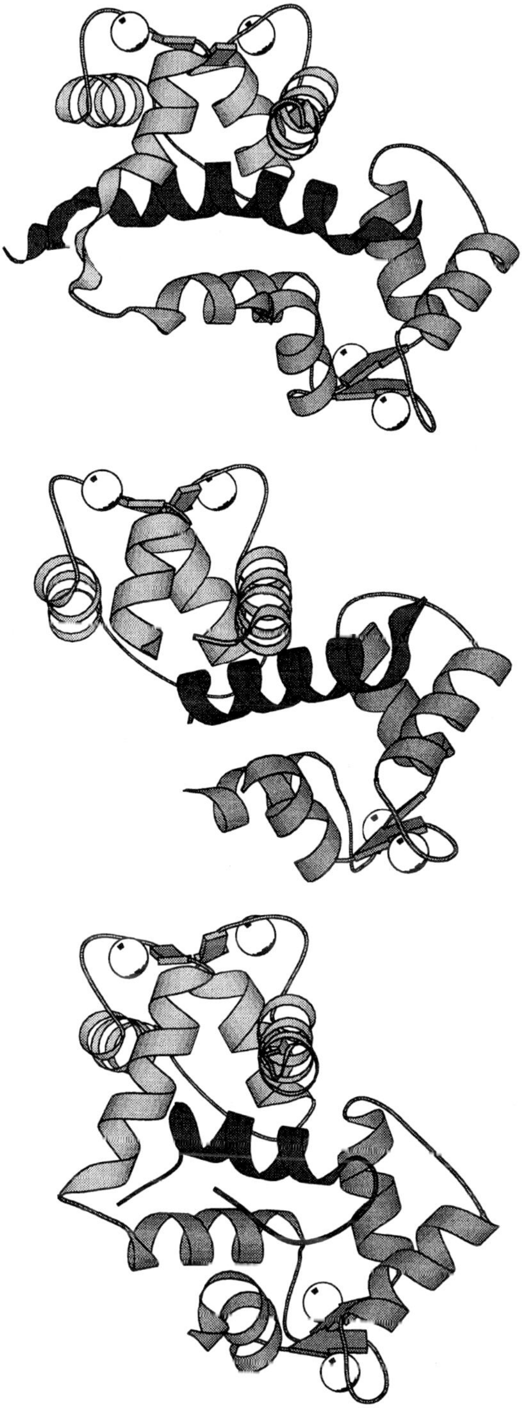

Fig. 5a–c. Structures of Ca^{2+}-CaM-peptide complexes: **a** Ca^{2+}-CaM-skMLCK-peptide complex [46]; **b** Ca^{2+}-CaM-CaMKII peptide complex [49]; **c** Ca^{2+}-CaM-CaMKK peptide complex [53]

two domains of Ca^{2+}-CaM wrap around the CaMKII peptide, which forms an α-helix in an antiparallel orientation similar to that observed in the CaM-MLCK complexes [46, 48]. CaMKII uses two bulky hydrophobic residues, L299 and L308, to anchor itself to the hydrophobic cavity of the C- and N-terminal domain of CaM, respectively (Fig. 4, 5B, 6B). These key hydrophobic residues are separated by 8 residues, which is four residues shorter than that of the MLCK interaction (Fig. 4). The central linker region of Ca^{2+}-CaM unwinds as seen in the CaM-MLCK peptide complex (Fig. 5B). The region between residues 73 and 83 has little electron density, presumably due to the flexibility. The shorter distance between the two anchoring residues in the CaMKII peptide indicates that the two domains of CaM are positioned closer to each other in this complex than in the CaM-MLCK peptide complex. This complex structure and the CaM-MLCK peptide complex structures demonstrate that the central linker region is crucial for CaM to recognize target proteins with different orientations optimal for each target protein (Fig. 5A, B).

4.3 Ca^{2+}-Calmodulin-CaMKK Complex

Given the approximate two-fold symmetry of the two domains of Ca^{2+}-CaM in complex with target peptides [46, 48, 49], both parallel and antiparallel orientations are possible for a peptide to bind to two CaM domains. A parallel orientation between CaM and a target peptide was first demonstrated in the Ca^{2+}-CaM-melittin complex [51, 52]. More recently, Osawa et al. [53] found a parallel binding orientation in the NMR structure of CaM bound to a CaM binding domain derived from CaM-dependent protein kinase kinase (CaMKK). This enzyme is an upstream activator of CaM-dependent protein kinase I (CaMKI) and CaM-dependent protein kinase IV (CaMKIV) as well as protein kinase B (for a recent review, see [54]). In this structure, the N-terminal half of a CaMKK peptide binds mainly to the N-terminal domain of CaM and the C-terminal half of the peptide to the C-terminal domain of CaM (Fig. 5C, 6C). There are two negatively charged outlets of the channel formed by the two CaM domains (Fig. 6C). Osawa et al. [53] noted that one channel outlet consists of more negative charges than the other channel outlet (Fig. 6C). It has been proposed that the electrostatic polarity of the channel is one important determinant for the parallel orientation between CaM and the CaMKK peptide: the positively-charged stretch (R455, K456, R457) at the C-terminal half of the CaMKK peptide preferentially binds the more negatively charged side of the binding channel (Fig. 6C). The significance of electrostatic interactions in CaM-target recognition was also noted by Meador et al. [49].

Another interesting feature of this structure is that, unlike the MLCK and CaMKII peptides, the CaMKK peptide assumes a hairpin-like loop (R455 to F459) in addition to an α-helix (W444 to L454), and both structural units are essential to the interaction with CaM (Fig. 5C, 6C). This C-terminal hairpin loop is also proposed to contribute to the orientation of the peptide, by virtue of asymmetric steric constraints of the peptide.

This structure represents the third class of CaM-target recognition, possessing the unique feature in secondary structure as well as a 14-residue separation between key hydrophobic residues. The N-terminal hydrophobic residue (W444) is located in the

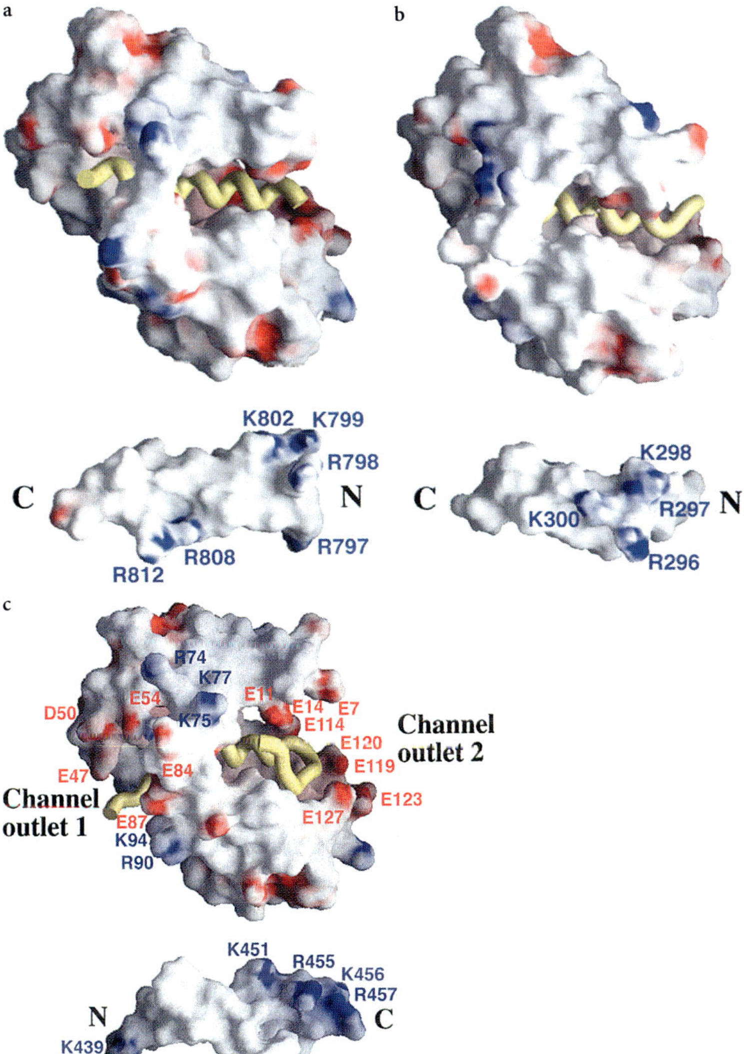

Fig. 6a–c. Electrostatic potential surfaces of CaM-target complexes: **a** Ca^{2+}-CaM-skMLCK-peptide complex [46]; **b** Ca^{2+}-CaM-CaMKII peptide complex [49]; **c** Ca^{2+}-CaM-CaMKK peptide complex [53]. In the *upper panels*, the surface of Ca^{2+}-CaM is shown with the target peptide as a *yellow tube*. The peptide surface is shown in the *lower panels*. The surface is colored according to the local electrostatic potential, with *blue* and *red* representing *positive* and *negative* potential, respectively. From [53]

helix, while the C-terminal hydrophobic residue (F459) resides in the hairpin loop region (Fig. 5). This separation of 16 residues is the longest "spacer" so far identified among the complex structures (Fig. 4). It is also interesting to note that the last residue of the peptide (F463) folds back onto the helix and interacts with the central linker region of CaM (M72 and M76). Similar to other CaM-peptide complexes, the central linker region of Ca^{2+}-CaM unwinds between residues 75 and 82 to accommodate the CaMKK peptide.

4.4 Ca^{2+}-CaM-Ca^{2+} Pump Complex

Elshorst et al. [55] recently reported a solution structure of Ca^{2+}-CaM complexed with the peptide C20W, a CaM binding site from the plasma membrane calcium pump. In this structure, C20W interacts only with the C-terminal domain of CaM, thereby allowing CaM to show an elongated shape with the two domains not interacting with each other. This is consistent with earlier SAXS studies, in which the gyration radius remains unchanged [56]. Note that C20W lacks the second anchoring residue V1120. In contrast, the C24W peptide, a peptide containing the full CaM binding sequence, produces a compact structure in the complex, as observed in the Ca^{2+}-CaM-MLCK peptide complex [46, 48, 56, 57]. As suggested by Elshorst et al. [55], the intimate interactions between the C-terminal domain of CaM and the C20W peptide probably account for the ability of the C-terminal domain to activate the Ca^{2+}-pump.

4.5 Calmodulin-Binding Domains in Target Proteins

To date, all NMR and X-ray structures of CaM-target complexes have used synthetic peptides corresponding to the CaM binding region of target proteins. Unfortunately, no structure is available for CaM complexed with an entire target protein or a domain containing both the catalytic core and a CaM binding domain. To date, four X-ray structures of target proteins containing CaM binding domains have been determined in the absence of CaM.

In 1996, Goldberg et al. [60] solved the crystal structure of CaMKI in the autoinhibited state. The regulatory segment of CaMKI contains a helix-loop-helix structure (αR1-loop-αR2) at the C-terminus (Fig. 7). The first α-helix and the loop region (αR1-loop) constitute the autoinhibitory domain, which interacts intimately with the catalytic core of the enzyme. The CaM binding domain contains the loop and second α-helix, both of which have extensive interactions with the ATP-binding region of the catalytic core. Interestingly, one of the potential anchoring residues in the CaM binding domain (W303) is fully exposed on the protein surface, while the other (M316) is completely buried inside the molecule (Fig. 7). Goldberg et al. [60] proposed that W303 provides an initial contacting site for CaM binding, and this tryptophan indeed interacts with the hydrophobic cavity in the C-terminal domain of Ca^{2+}-CaM [61]. This is consistent with a proposed activation process in which the interaction between W303 and the C-terminal domain of Ca^{2+}-CaM initiates the detachment of the inhibitory domain from the catalytic core [61, 62].

Another protein containing a CaM binding domain of which a crystal structure has been solved is titin kinase [63]. Similar to CaMKI, the regulatory segment comprises αR1-loop-αR2 (Fig. 7). The CaM binding domain (loop-αR2) also has

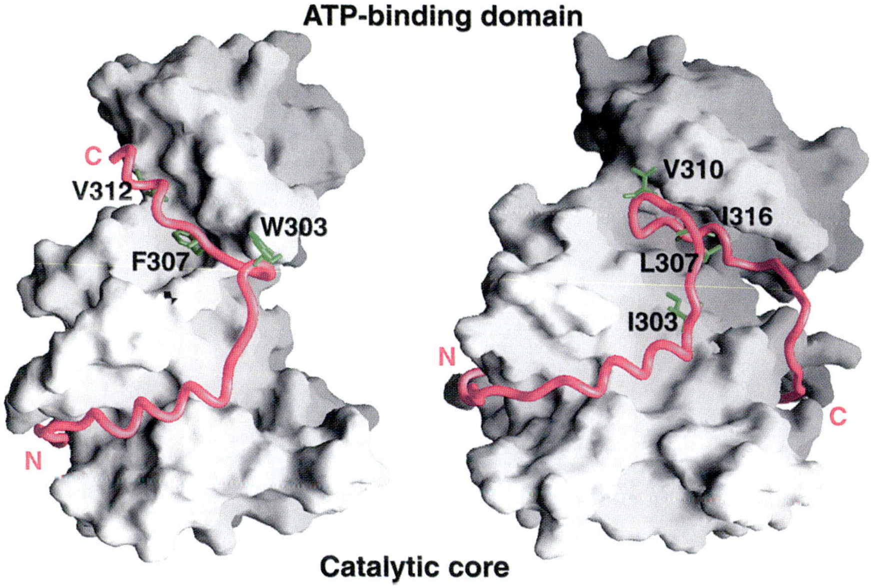

Fig. 7. The structures of CaMKI [60] and titin kinase [63] in the autoinhibited state. The catalytic domain is shown as a surface representation. The regulatory region containing the autoinhibitory and CaM-binding domains are shown as a *magenta tube*. Hydrophobic residues essential for CaM binding are *highlighted in green*

many interactions with the kinase's ATP-binding sites. However, both potential anchoring residues in this CaM binding sequence (I286 and I299) are not fully exposed. This suggests that CaM binds to titin kinase in a manner different from binding to MLCK or CaMKI. Alternatively, the anchoring residues in its CaM binding sequence may not be I286 and I299.

In the structures of calcineurin and type II adenylyl cyclase [58, 59], the CaM binding domains are invisible in the electron density map due to their weak intensity. This potentially flexible nature of the CaM binding domain within the proteins may play a role in the recruitment of Ca^{2+}-CaM for activation.

5 Structures of Calmodulin-Drug Complexes

5.1 Ca^{2+}-Calmodulin-TFP Complex

A number of inhibitors have been found for CaM from many biological and pharmacological studies over the last two decades [64, 65]. TFP (trifluoperazine, 10-[3-(4-methylpiperazin-1-yl)-propyl]-2-trifluoromethylphenothiazine), one of the most potent antipsychotic phenothioazine-type drugs, binds and inhibits CaM. The dissociation constant of TFP for Ca^{2+}-CaM is 1 μM and the binding stoichiometry ranges from two to seven molecules of TFP bound to one Ca^{2+}-CaM molecule, depending on solution conditions (summarized in [66]). A 2.0-Å X-ray structure of Ca^{2+}-CaM bound to four TFP molecules was determined in 1994 [66]. TFP1 and TFP4 mainly interact with the hydrophobic cavities in the C- and N-terminal domains of Ca^{2+}-CaM, respectively. TFP2 and TFP3 bind to a region between the two domains of CaM with their phenothiazine rings interacting with both hydrophobic cavities of CaM, leaving the piperazine rings protruding outside to interact electrostatically with acidic residues on the surface of CaM. Numerous contacts between TFP molecules and the hydrophobic cavities of Ca^{2+}-CaM prevent CaM from binding its target proteins, providing a structural basis for TFP's inhibition of CaM's various functions. The central linker region between residues 75 and 80 becomes unwound in the Ca^{2+}-CaM-TFP structure such that the global shape resembles the structures of Ca^{2+}-CaM-peptide complexes [46, 48, 49, 53].

As mentioned above, the stoichiometry of TFP binding to Ca^{2+}-CaM is variable depending on experimental conditions. This phenomenon was observed in two additional Ca^{2+}-CaM-TFP complexes [67, 68]. Vertessy et al. [68] found that two TFP molecules (TFP1 and TFP2) bind to the hydrophobic cavity of the C-terminal domain of CaM and an inter-domain site, respectively, similar to TFP1 and TFP2 in the structure of Vandonselaar et al. [66]. This Ca^{2+}-CaM-TFP complex also has a globular shape with a bending between residues 74 and 80 [68]. Surprisingly, Cook et al. [67] found that only one TFP molecule binds to the hydrophobic cavity of the C-terminal domain of CaM, and the tricyclic ring of TFP is rotated 180° compared with the other two structures. The structure of Cook et al. [67] adopts a globular shape with a sharp bend between residues 75 and 80 in CaM. Differences observed in the above three X-ray structures of the CaM-TFP complex probably originate from the crystallization conditions used, such as pH and molecular ratio of TFP to CaM [67, 68].

5.2 Ca^{2+}-Calmodulin-W-7 Complex

N-(6-Aminohexyl)-5-chloro-1-naphthalene sulfonamide, or W-7, is a well-known compound used for studies to investigate CaM-regulated signal transduction pathways. Its dissociation constant for Ca^{2+}-CaM is around 11 μmol/l [65]. A solution structure of Ca^{2+}-CaM complexed with two W-7 molecules was determined by multidimensional NMR spectroscopy [69] (Fig. 8). The 5-chloro-1-naphthalene ring of each W-7 molecule is inserted into the hydrophobic cavity of each CaM domain with the chlorine atom pointing toward the base of the cavity. Most interestingly, the side chains of four Met residues in each domain of CaM undergo a fine readjustment in their conformations to accommodate the naphthalene ring of W-7. The aromatic residues F19 and F92 also play an important role in defining the size of each W-7 binding pocket. Although the nOe based structural determination did not yield constraints between the two domains of CaM in the complex, recent SAXS studies indicated a compaction of Ca^{2+}-CaM upon W-7 binding [70], similar to the result obtained for the Ca^{2+}-CaM-TFP complex [66–68]. These results demonstrate that binding of inhibitors to the hydrophobic cavity itself influences the orientation of the N- and C-terminal domains of CaM.

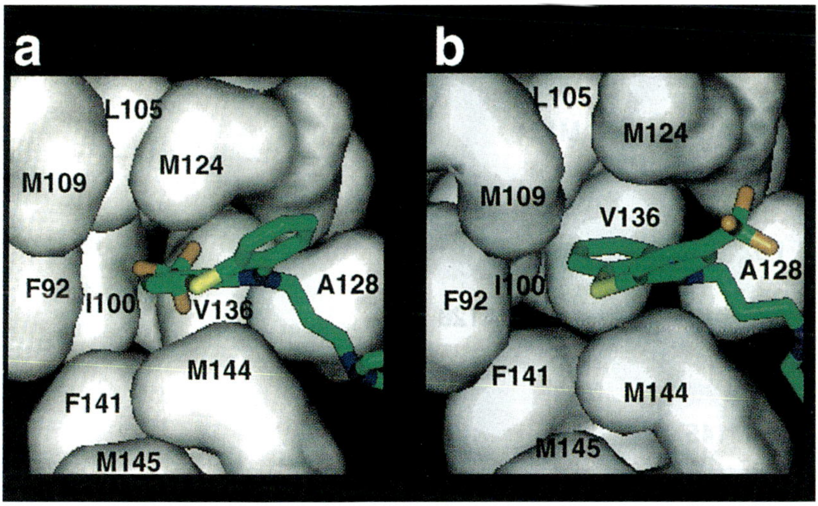

Fig. 8a, b. The hydrophobic cavity of CaM C-terminal domain bound to: **a** TFP; **b** W-7. From [69]

6
Other Calmodulin Binding Motifs

The CaM binding motifs discussed so far all contain a single α-helix. In the following sections we will discuss three CaM binding proteins, which have been shown to possess unique structural and functional features. Many other CaM binding domains are reviewed in Crivici and Ikura [3], Rhoads and Friedberg [5], and Zhang and Yuan [11].

6.1
MARCKS and Related Proteins

Myristoylated alanine-rich C kinase substrate (MARCKS) is a substrate for protein kinase C, and has been proposed to regulate the actin cytoskeleton [71]. MARCKS contains a stretch of basic and hydrophobic residues which binds to Ca^{2+}-CaM with a high affinity (~1 nmol/l). Biophysical studies using circular dichroism (CD) and NMR spectroscopy suggested that this CaM binding domain does not assume an α-helical conformation, but rather adopts an extended conformation upon binding to Ca^{2+}-CaM [72, 73]. Similarly, two other CaM binding domains bind to Ca^{2+}-CaM in a non-α-helical conformation: MARCKS-related protein (F52/MRP) and Ras-related GTP-binding protein (Ral-A) [74, 75]. Notably, positively charged residues in these CaM binding domains constitute more than 50% of the entire binding site (summarized in [73]) (Fig. 9A). Some CaM binding proteins share the same property, including human diacylglycerol kinase ζ, α-adducin, β-adducin and 35H [76–79]. No high-resolution structural data have been obtained for a CaM binding protein of this type nor for a derived peptide bound to CaM.

(A)

```
MARCKS   152- K K K K K R F S F K K S F K L S G F S F K K N K K -176
F52/MRP        88- K K K K F S F K K P F K L S G L S F K R N R K -110

Ral-A        183- S K E K N G K K K R K S L A K R I R -200
```

(B)

```
IQ motif                           I Q X X X R G X X X R

L-type (α1C)          1649- Y A T F L I Q E Y F R K F K K R K E Q -1667
P/Q-type (α1A)        1908- Y A A M M I M E Y Y R Q S K A K K L Q -1926
R-type (α1E)          1821- Y A A M M I M D Y Y K Q S K V K K Q R -1839
```

(C)

```
PGD     470- H K K T D S E V Q L E M I T A W K K F V E E K K K K -495
```

Fig. 9A–C. Amino acid sequences of other CaM binding motifs: **A** CaM binding peptides which may not adopt an α-helical structure upon binding to CaM; **B** the IQ-motif found in Ca^{2+}-channels. These binding sites bind to CaM both in the presence and absence of calcium. The consensus amino acid sequence of the IQ motif is indicated above the sequences; **C** CaM binding domain from petunia glutamate decarboxylase. Two molecules of the peptide bind to one molecule of Ca^{2+}-CaM simultaneously [92]

6.2 Ion Channels

A number of ion channels appear to be regulated by Ca^{2+}-CaM (Fig. 9B). These ion channels include *N*-methyl-D-aspartate (NMDA) receptor [80], olfactory cyclic nucleotide-gated (CNG) ion channels [81], small conductance (SK) calcium-activated potassium channels [82, 83], intermediate conductance (IK) calcium-activated potassium channels [84], transient receptor potential-like (TRPL) ion channels [85], L-type calcium channels [86–88], P/Q-type calcium channels [86, 89] and R-type calcium channels [86]. A unique feature of these CaM binding ion channels is that many of them bind to CaM constitutively in a Ca^{2+}-independent manner. This constitutive binding of CaM involves two CaM binding sites in some cases [81, 85]. Of particular interest is a study reported by Zuhlke et al. [88], in which CaM binding to an IQ-motif in the C-terminus of α_{1c} subunit of an L-type Ca^{2+} channel was found to support both inactivation and facilitation of the channel. CaM facilitates the channel activity at the resting concentration of calcium, while it inactivates the channel at a high Ca^{2+} concentration. Zuhlke et al. [88] showed that mutation of I1624 in the IQ-motif to Ala abolishes Ca^{2+}-dependent inactivation and strongly facilitates the activation process, while mutation of this Ile to Glu eliminates both functions. Thus CaM plays a unique role in the dual regulation of the L-type calcium channel, supporting a local, rapid, and feedback response to the Ca^{2+} concentration fluctuation [88]. Regions other than the IQ-motif of the L-type calcium channel

[86, 87] and other ion channels [82, 83] may also be involved in Ca^{2+}-independent tethering of CaM.

6.3
Glutamate Decarboxylase

Plant glutamate decarboxylase (PGD) is a unique enzyme which catalyzes the conversion of glutamate to γ-butyrate (GABA) during differentiation and development [90]. This plant enzyme is unique in that it contains ~25 amino acid residues at the C-terminus possessing high CaM-binding activity [91], and the binding site is absent in PGD found in mammals and *E. coli.* This CaM binding site is unusual because it contains five negatively charged residues in addition to basic and hydrophobic residues that are usually found in CaM-binding motifs (Fig. 9C). Surprisingly, binding of Ca^{2+}-CaM induces the dimer formation of the enzyme. It has been shown [92] that even a peptide encoding the 25-residue CaM binding site forms a dimer when bound to Ca^{2+}-CaM [92]. Preliminary NMR studies indicate that the two peptide molecules interact with each other while binding cooperatively to Ca^{2+}-CaM (Yuan, Vogel and Ikura, unpublished observations).

7
Structural Basis of Calmodulin Functional Versatility

More than 100 CaM-binding proteins have been discovered in the last two decades (compiled in the Calmodulin Target Database, at http://calcium.oci.utoronto.ca). High-resolution structures of CaM in different forms, namely apo-CaM, Ca^{2+}-CaM and the Ca^{2+}-CaM-peptide complexes, greatly enrich our understanding of the target recognition process by CaM. As described above, there are at least three different classes of CaM-target recognition: the 1–14 class, represented by CaM-MLCK peptide complexes [46, 48], the 1–10 class, represented by the CaM-CaMKII peptide complex [49] and the 1–16 class, represented by the CaM-CaMKK peptide complex [53]. In addition, the IQ motif is likely to possess a unique structure when bound to apo-CaM. All the known structures indicate the significance of several structural properties of CaM in its multi-target recognition. Below we discuss three key structural features in CaM that are clearly important for its unique target recognition activities.

7.1
Flexible Domain Linker

The central linker region of CaM comprises the second α-helix of EF-hand II, a linker and the first α-helix of EF-hand III (Fig. 1). In the crystal structure of Ca^{2+}-CaM, the central linker region forms a continuous α-helix. However, solution studies using various techniques such as limited protease digestion, chemical cross-linking, mutagenesis, SAXS, NMR and molecular simulation clearly demonstrated that the middle portion of the central helix (residues 78–81) does not assume an α-helix and experiences high flexibility in its polypeptide conformation [19, 20, 23–26, 93–96].

The X-ray structure of a CaM mutant lacking E84 showed a 95° bend in the central linker region, implying the intrinsic flexibility of this region [97]. Tabernero et al. [98] solved a high-resolution (1.8 Å) X-ray structure of a mutant Ca^{2+}-CaM lacking both T79 and D80 in the central linker region. This deletion causes a rotation of the C-terminal domain of CaM about 220° relative to the N-terminal domain. As a consequence, the two domains of CaM are in a *cis* rather than a *trans* position observed in the X-ray structure of wild type CaM (WT-CaM) (Fig. 2). Two hydrophobic cavities are positioned parallel to each other and are less accessible to its target proteins. Indeed, the activation of phosphodiesterase (PDE) was reduced in this CaM mutant, with a level 1/15 that of WT-CaM [94]. In contrast, this CaM mutant is able to bind to a CaMKII peptide through the adjustment of the uncoiled central linker region [99]. Furthermore, the central linker region of CaM can remain extended in order to bind to a target sequence in phosphorylase kinase, as revealed by SAXS studies [100]. These data highlight the key role of the flexible central linker for CaM to recognize various target proteins with different binding sequences.

7.2 Methionine-Rich Hydrophobic Cavities

The content of Met residues of CaM is extremely high (6%) compared to the average content of all known proteins (1.5%) [7]. Eight out of nine Met residues in CaM are found in the hydrophobic cavities responsible for target binding. The Met residues are largely or partially buried in the N- and C-terminal domain of apo-CaM, respectively [31, 43]. Upon Ca^{2+}-binding, all the Met residues become fully exposed, and 46% of the hydrophobic surface of the two hydrophobic cavities is occupied by these Met residues [7, 14]. In the structures of Ca^{2+}-CaM-peptide and Ca^{2+}-CaM-drug complexes, almost all Met residues are involved in hydrophobic interactions with CaM's targets [46, 48, 49, 53, 55, 66–69, 99].

Being an unbranched hydrophobic, sulfur-containing residue, methionine has two unique properties [47]. First, the Met side chain is very flexible because of the low energy barrier between *anti* and *gauche* rotamers at the C—S bond. Second, the sulfur atom is highly polarizable compared to the methylene group, and thus the hydrophobic surface created by Met residues is malleable. Indeed, site-directed mutagenesis, incorporation of unnatural amino acid analogs of Met and chemical modification studies all indicate that Met residues of CaM are indispensable for CaM's versatility [101–108]. Substitution of Met residues in CaM with norleucine, with a methylene group replacing the sulfur atom, also impairs CaM's ability to activate its target proteins [106]. In contrast, replacement of the sulfur atom within Met residues with a more polarizable selenium atom fully restores CaM's function [105]. These studies illustrate the importance of the unique properties of Met residues in CaM. Furthermore, the structural plasticity of the Met side chains was demonstrated in the structure of the CaM-W-7 complex [69]. In this structure, the biggest chemical shift changes among all methyl groups in CaM are assigned to the Met methyl groups, consistent with the plasticity of the Met side chains in the two hydrophobic cavities of Ca^{2+}-CaM. Similar observations are presented by Elshorst et al. [55] for the Ca^{2+}-CaM-C20W peptide complex. Depending on the target protein, CaM uses a particular set of Met side chains to achieve the most stable complex formation [43]. Clearly, the Met residues in the two hydrophobic cavities of Ca^{2+}-CaM are indispensable for its multiple target binding functionality.

7.3 Ca^{2+}-Independent Target Binding

A handful of targets are known to bind CaM in the absence of Ca^{2+}. A list of Ca^{2+}-independent CaM binding motifs has been compiled by Rhoads and Friedberg [5]. A possible function of apo-CaM-target interaction is either to sequester CaM close to the membrane in the case of neuromodulin and neurogranin, or to bind CaM constitutively in the case of ion channels as described above (see Sect. 6.2). Due to lack of a high-resolution structure of apo-CaM bound to a target, the binding mechanism is still largely unknown at present. However, the structure of apo-CaM shows scattered hydrophobic residues and many negatively charged residues on the surface [31]. In accordance with this observation, CaM binding sites in target proteins contain positively charged and hydrophobic residues [5, 109]. As discussed in Sect. 3, the C-terminal domain of apo-CaM probably provides a primary binding site for its targets. Its dynamic nature and partially exposed hydrophobic surface are likely the signature of the binding site [30, 31, 42, 110, 111]. This was confirmed by many biophysical studies [39, 44, 112]. Upon target peptide binding, the most significant perturbations of NMR resonance of apo-CaM are located on the C-terminal domain [39–44]. The ability of CaM to bind to its targets in the absence of Ca^{2+} extends its structural and functional versatility.

8 Future Directions

Since 1985, more than a dozen high-resolution structures have become available for CaM and its complexes with target peptides and drugs. These structures have greatly facilitated our understanding of CaM's structures and functions. Current directions in the field involve the following tasks.

8.1 Determination of a High-Resolution Structure of a Complex Between Ca^{2+}-CaM and Its Target Protein or a Large Protein Fragment

Although the CaM binding sequence is usually confined to a ~25-residue peptide in the target protein, there is increasing evidence of additional binding sites between CaM and its target proteins. The canonical CaM binding sequence is the primary binding site for CaM, and any other interactions are regarded as "secondary interaction" sites. Persechini et al. [113] reported an interesting case in which the dele tion of the N-terminal sequence Glu6-Glu7-Gln8 in CaM significantly impairs its ability to activate MLCK. However, this portion of CaM has little contact with the CaM binding domain peptide from MLCK [46, 48]. Zhi et al. [114] made several chimeric kinase mutants using the catalytic core of skMLCK or CaMKII, and the regulatory domain of skMLCK, smMLCK or CaMKII. They found that the activities of the chimeric kinases are sensitive to the catalytic core instead of the CaM-binding domain sequence. The concentrations of CaM for half maximum activation (K_{CaM}) of CaMKII and chimeric kinases containing the CaMKII catalytic core are all similar ($25 \sim 100$ nmol/l), regardless of whether the CaM-binding sequence is from CaMKII, skMLCK or smMLCK [114]. Similar results were observed for MLCK chimeric proteins ($K_{CaM} \sim 1$ nmol/l). These results suggest secondary interaction between CaM and the catalytic core of MLCK or CaMKII, which may be crucial to the activation of its targets. Further determination of structures of the CaM-MLCK protein complex or the CaM-CaMKII protein complex will shed more light on our understanding of the nature of this secondary interaction. Such structural studies have already been initiated by Trewhella's group using SAXS techniques [115, 116]. Similarly, secondary interaction between CaM and its target proteins has also been reported for inducible nitric oxide synthase [117, 118].

8.2 Determination of a High-Resolution Structure of an apo-CaM-Target Complex

Although NMR structures of apo-CaM are available [29–31] and there is a detailed study of apo-CaM-peptide interactions [39], it is still unclear how apo-CaM interacts with its target proteins. The intrinsic dynamic nature of the C-terminal domain of CaM may pose a challenge in obtaining a good crystal or NMR sample for structure determination. Nevertheless, attempts have been and are being made in the right direction. For example, a crystal of an apo-CaM-drug complex was reported by Vertessy et al. [119].

8.3 Continuation of CaM-Target Peptide Complex Studies

More CaM-target complex structures need to be determined in order to understand fully CaM's multiple target recognition function. Some unique CaM-peptide complexes are currently under investigation, such as those discussed in Sect. 6, as well as many more yet to come. The Calmodulin Target Database should serve as a forum in which all known structures and sequences containing CaM binding motifs are summarized.

References

1. Kretsinger RH, Nockolds CE (1973) J Biol Chem 248:3313
2. Copley RR, Schultz J, Ponting CP, Bork P (1999) Curr Opin Struct Biol 9:408
3. Crivici A, Ikura M (1995) Ann Rev Biophys Biomol Struct 24:85
4. Vogel HJ, Zhang M (1995) Mol Cell Biochem 149/150:3
5. Rhoads AR, Friedberg F (1997) FASEB J 11:331
6. Means AR, VanBerkum MFA, Bagchi I, Lu KP, Rasmussen CD (1991) Pharmacol Ther 50:255
7. O'Neil KT, DeGrado WF (1990) Trends Biochem Sci 15:59
8. Vogel HJ (1994) Biochem Cell Biol 72:357
9. James P, Vorherr T, Carafoli E (1995) Trends Biochem Sci 20:38

10. Ikura M (1996) Trends Biochem Sci 21:14
11. Zhang M, Yuan T (1998) Biochem Cell Biol 76:313
12. Babu YS, Sack JS, Greenhough TJ, Bugg CE, Means AR, Cook WJ (1985) Nature 315:37
13. Kretsinger RH, Rudnick S, Weissman LJ (1986) J Inorg Biochemistry 28:289
14. Babu YS, Bugg CE, Cook WJ (1988) J Mol Biol 204:191
15. Chattopadhyaya R, Meador WE, Means AR, Quiocho FA (1992) J Mol Biol 228:1177
16. Taylor DA, Sack JS, Maune JF, Beckingham K, Quiocho FA (1991) J Biol Chem 266:21,375
17. Rao ST, Wu S, Satyshur KA, Ling K-Y, Kung C, Sundaralingam M (1993) Prot Sci 2:436
18. Ban C, Ramakrishnan B, Ling K-Y, Kung C, Sundaralingam M (1994) Acta Cryst D50:50
19. Walsh M, Stevens FC, Kuznicki J, Drabikowski W (1977) J Biol Chem 252:7440
20. Drabikowski W, Kuznicki J, Grabarek Z (1977) Biochim Biophys Acta 485:124
21. LaPorte DC, Wierman BM, Storm DR (1980) Biochemistry 19:3814
22. Tanaka T, Hidaka H (1980) J Biol Chem 255:11,078
23. Heidorn DB, Trewhella J (1988) Biochemistry 27:909
24. Ikura M, Spera S, Barbato G, Kay LE, Krinks M, Bax A (1991) Biochemistry 30:9216
25. Spera S, Ikura M, Bax A (1991) J BioMol NMR 1:155
26. Barbato G, Ikura M, Kay LE, Pastor RW, Bax A (1992) Biochemistry 31:5269
27. Bayley PM, Martin SR (1992) Biochim Biophys Acta 1160:16
28. Herzberg O, Moult J, James MN (1986) J Biol Chem 261:2638
29. Finn BE, Evenas J, Drakenberg T, Waltho JP, Thulin E, Forsen S (1995) Nat Struct Biol 2:777
30. Kuboniwa H, Tjandra N, Grzesiek S, Ren H, Klee CB, Bax A (1995) Nat Struct Biol 2:768
31. Zhang M, Tanaka T, Ikura M (1995) Nat Struct Biol 2:758
32. Travo G, Lacombe PJ, Pfuhl M, Saraste M, Pastore A (1995) EMBO J 14:4922
33. Tanaka T, Ames JB, Harvey TS, Stryer L, Ikura M (1995) Nature 376:444
34. Ames JB, Ishima R, Tanaka T, Gordon JI, Stryer L, Ikura M (1997) Nature 389:198
35. Gagne SM, Li MX, McKay RT, Sykes BD (1998) Biochem Cell Biol 76:302
36. Smith SP, Shaw GS (1998) Biochem Cell Biol 76:324
37. Yap KL, Ames JB, Swindells MB, Ikura M (1999) Proteins 37:499
38. Skelton NJ, Kördel J, Akke M, Forsén S, Chazin WJ (1994) Nat Struct Biol 1:239
39. Urbauer JL, Short JH, Dow LK, Wand AJ (1995) Biochemistry 34:8099
40. Tjandra N, Kuboniwa H, Ren H, Bax A (1995) Eur J Biochem 230:1014
41. Tsalkova TN, Privalov P (1985) J Mol Biol 181:533
42. Swindells MB, Ikura M (1996) Nat Struct Biol 3:501
43. Yuan T, Ouyang H, Vogel HJ (1999) J Biol Chem 274:8411
44. Yuan T, Walsh MP, Sutherland C, Fabian H, Vogel HJ (1999) Biochemistry 38:1446
45. Stull JT, Kamm KE, Krueger JK, Lin P, Luby-Phelps K, Zhi G (1997) Adv Second Messenger Phosphoprotein Res 31:141
46. Ikura M, Clore GM, Gronenborn AM, Zhu G, Klee CB, Bax A (1992) Science 256:632
47. Gellman SH (1991) Biochemistry 30:6633
48. Meador WE, Means AR, Quiocho FA (1992) Science 257:1251
49. Meador WE, Means AR, Quiocho FA (1993) Science 262:1718
50. Braun AP, Schulman H (1995) Ann Rev Physiol 57:417
51. Quiocho FA, Johnson KA, Meador WE, Beckingham K (1997) Proceedings of the 10th International Symposium on Calcium-Binding Proteins and Calcium Function in Health and Disease, Lund, Sweden, p 27
52. Scaloni A, Miraglia N, Orru S, Amodeo P, Motta A, Marino G, Pucci P (1998) J Mol Biol 277:945
53. Osawa M, Tokumitsu H, Swindells MB, Kurihara H, Orita M, Shibanuma T, Furuya T, Ikura M (1999) Nat Struct Biol 6:819
54. Soderling TR (1999) Trends Biochem Sci 24:232
55. Elshorst B, Hennig M, Forsterling H, Diener A, Maurer M, Schulte P, Schwalbe H, Griesinger C, Krebs J, Schmid H, Vorherr T, Carafoli E (1999) Biochemistry 38:12,320
56. Kataoka M, Head JF, Vorherr T, Krebs J, Carafoli E (1991) Biochemistry 30:6247
57. Heidorn DB, Seeger PA, Rokop SE, Blumenthal DK, Means AR, Crespi H, Trewhella J (1989) Biochemistry 28:6757
58. Kissinger CR, Parge HE, Knighton DR, Lewis CT, Pelletier LA, Tempczyk A, Kalish VJ, Tucker KD, Showalter RE, Moomaw EW, Gastinel LN, Habuka N, Chen X, Maldonado F, Barker JE, Bacquet R, Villafranca JE (1995) Nature 378:641

59. Zhang G, Liu Y, Ruoho AE, Hurley JH (1997) Nature 386:247
60. Goldberg J, Nairn AC, Kuriyan J (1996) Cell 84:875
61. Yuan Y, Weljie AM, Vogel HJ (1998) Biochemistry 37:3187
62. Bayley PM, Findlay WA, Martin SR (1996) Prot Sci 5:1215
63. Mayans O, van der Ven PFM, Wilm M, Mues A, Young P, Furst DO, Wilmanns M, Gautel M (1998) Nature 395:863
64. Levin RM, Weiss B (1977) Mol Pharmacol 13:690
65. Hidaka H, Yamaki T, Naka M, Tanaka T, Hayashi H, Kobayashi R (1979) Mol Pharmacol 17:66
66. Vandonselaar M, Hickie RA, Quail JW, Delbaere LTJ (1994) Nat Struct Biol 1:795
67. Cook WJ, Walter LJ, Walter MR (1994) Biochemistry 33:15,259
68. Vertessy BG, Harmat V, Bocskei Z, Naray-Szabo G, Orosz F, Ovadi J (1998) Biochemistry 37:15,300
69. Osawa M, Swindells MB, Tanikawa J, Tanaka T, Mase T, Furuya T, Ikura M (1998) J Mol Biol 276:165
70. Osawa M, Kuwamoto S, Izumi Y, Yap KL, Ikura M, Shibanuma T, Yokokura H, Hidaka H, Matsushima N (1999) FEBS Lett 442:173
71. Blackshear PJ (1993) J Biol Chem 268:1501
72. Porumb T, Crivici A, Blackshear PJ, Ikura M (1997) Eur Biophys J 25:239
73. Matsubara M, Yamauchi E, Hayashi N, Taniguchi H (1998) FEBS Lett 421:203
74. Schleiff E, Schmitz A, McIlhinney RA, Manenti S, Vergeres G (1996) J Biol Chem 271:26,794
75. Wang KL, Khan MT, Roufogalis BD (1997) J Biol Chem 272:16,002
76. Bunting MD, Zhang S, Bernhardt JP, Huganir RL (1996) J Biol Chem 271:10,230
77. Matsuoka Y, Hughes CA, Bennett V (1996) J Biol Chem 271:25,157
78. Joshi R, Bennett V (1990) J Biol Chem 265:13,130
79. Dong L, Chapline C, Mousseau B, Fowler L, Ramsay K, Stevens JL, Jaken S (1995) J Biol Chem 270:25,534
80. Ehlers MD, Zhang S, Bernhardt JP, Huganir RL (1996) Cell 84:745
81. Grunwald ME, Yu W-P, Yu H-H, Yau K-W (1998) J Biol Chem 273:9148
82. Xia X-M, Fakler B, Rivard A, Wayman G, Johnson-Pais T, Keen JE, Ishii T, Hirschberg B, Bond CT, Lutsenko S, Maylie J, Adelman JP (1998) Nature 395:503
83. Khanna R, Chang MC, Joiner WJ, Kaczmarek LK, Schlichter LC (1999) J Biol Chem 274:14,838
84. Fanger CM, Ghanshani S, Logsdon NJ, Rauer H, Kalman K, Zhou J, Beckingham K, Chandy KG, Cahalan MD, Aiyar J (1999) J Biol Chem 274:5746
85. Trost C, Marquart A, Zimmer S, Philipp S, Cavalie A, Flockerzi V (1999) FEBS Lett 451:257–263
86. Peterson BZ, DeMaria CD, Yue, DT (1999) Neuron 22:549
87. Qin N, Olcese R, Bransby M, Lin T, Birnbaumer L (1999) Proc Natl Acad Sci 96:2435
88. Zuhlke RD, Pitt GS, Deisseroth K, Tsien RW, Reuter H (1999) Nature 399:159
89. Lee A, Wong ST, Gallagher D, Li B, Storm DR, Scheuer T, Catterall WA (1999) Nature 399:155
90. Baum G, Lev-Yadun S, Fridmann Y, Arazi T, Katsnelson H, Zik M, Fromm H (1996) EMBO J 15:2988
91. Baum G, Chen Y, Arazi T, Takatsuji H, Fromm H (1993) J Biol Chem 268:19,610
92. Yuan T, Vogel HJ (1998) J Biol Chem 273:30,328
93. Persechini A, Kretsinger RH (1988) J Biol Chem 263:12,175
94. VanBerkum MFA, George SE, Means AR (1990) Calmodulin activation of target enzymes. Consequence of deletions in the central helix. J Biol Chem 265:3750–3756
95. van der Spoel D, DeGroot BL, Hayward S, Berendsen HJ, Vogel HJ (1996) Prot Sci 5:2044
96. Wriggers W, Mehler E, Pitici F, Weinstein H, Schulten K (1998) Biophys J 74:1622
97. Raghunathan S, Chandross RJ, Cheng B-P, Persechini A, Sobottka SE, Kretsinger RH (1993) Proc Natl Acad Sci 90:6869
98. Tabernero L, Taylor DA, Chandross R, VanBerkum MFA, Means AR, Quiocho FA, Sack JS (1997) Structure 5:613
99. Meador WE, George SE, Means AR, Quiocho FA (1995) Nat Struct Biol 2:943
100. Trewhella J, Blumenthal DK, Rokop SE, Seeger PA (1990) Biochemistry 29:9316
101. Zhang M, Li M, Wang JH, Vogel HJ (1994) J Biol Chem 269:15,546
102. Edwards RA, Walsh MP, Sutherland C, Vogel HJ (1998) Biochem J 331:149
103. Chin D, Means AR (1996) J Biol Chem 271:30,465
104. Chin D, Winkler KE, Means AR (1997) J Biol Chem 272:31,235

105. Zhang M, Vogel HJ (1994) J Mol Biol 239:545
106. Yuan T, Vogel HJ (1999) Prot Sci 8:113
107. Walsh MP, Stevens FC (1978) Biochemistry 17:3924
108. Yao Y, Yin D, Jas GS, Kuczer K, Williams TD, Schoneich C, Squier TC (1996) Biochemistry 35:2767
109. Zhu T, Ikebe MA (1994) FEBS Lett 339:31
110. Houdusse A, Cohen C (1995) Proc Natl Acad Sci USA 92:10,644
111. Houdusse A, Silver M, Cohen C (1996) Structure 4:1475
112. Tsvetkov PO, Protasevich II, Gill R, Lafitte D, Lobachov VM, Haiech J, Briand C, Makarov AA (1999) J Biol Chem 274:18,161
113. Persechini A, Gansz K, Paresi R (1996) J Biol Chem 271:19,279
114. Zhi G, Abdullah SM, Stull JT (1998) J Biol Chem 273:8951
115. Krueger J, Olah GA, Rokop SE, Zhi G, Stull JT, Trewhella J (1997) Biochemistry 36:6017
116. Krueger JK, Zhi G, Stull JT, Trewhella J (1998) Biochemistry 37:13,997
117. Ruan J, Xie Q-W, Hutchinson N, Cho H, Wolfe GC, Nathan C (1996) J Biol Chem 271:22,679
118. Lee S-J, Stull JT (1998) J Biol Chem 273:27,430
119. Vertessy BG, Bocskei Z, Harmath V, Naray-Szabo G, Ovadi J (1997) Proteins 28:131

Calcium-Binding EGF-like Domains

A.K. Downing, P.A. Handford, I.D. Campbell*

University of Oxford, Department of Biochemistry and Oxford Centre for Molecular Sciences, South Parks Road, Oxford, OX1 3QU, UK

The calcium-binding EGF-like (cbEGF) domain is a common module found in functionally diverse extracellular proteins. Its importance is highlighted by a number of inherited diseases caused by missense mutations within the domain. It is characterised by six cysteine residues which normally disulphide bond in a 1–3, 2–4, 5–6 pattern and a calcium-binding consensus D/N-x-D/N-E/Q-x_m-D/N*-x_n-Y/F (where m and n are variables and * indicates possible β-hydroxylation). High-resolution structures of cbEGF domains have identified a common fold comprising major and minor double-stranded β-hairpins. The domain lacks a significant hydrophobic core and the structure is stabilised by the three consensus disulphide bonds and calcium binding to the amino terminus. Calcium is coordinated by a pentagonal bipyramidal arrangement of oxygen atoms, six of which are provided by intradomain ligands. Each individual cbEGF domain contains a weak calcium-binding site, the affinity of which may be substantially enhanced by N-terminal linkage to another domain. Structural analysis of cbEGF domains linked either to homologous or heterologous domains identifies a general role for calcium in restricting the conformational flexibility of interdomain linkages and orienting domain pairs.

Keywords. Extracellular, Calcium, EGF, Protein, Module

1 Introduction

In this review we concentrate on the role of extracellular calcium ions in higher organisms and use the calcium-binding EGF module as our main illustrative example. The intracellular signalling role of calcium ions is well documented [1]. There are numerous intracellular Ca^{2+} binding proteins, most of them constructed from "EF hand" motifs [2, 3]. They are tuned to bind and respond to changes in intracellular Ca^{2+} concentrations in the range 10 nM–10 μM. Although less well documented, Ca^{2+} also has a key role in the extracellular space, where its concentration is tightly regulated [4]. The total Ca^{2+} concentration in serum is maintained, very close to 2.4 mM, by an endocrine system that includes parathyroid hormone, vitamin D and calcitonin. Errors in regulation caused, for example, by mutations in a calcium-sensing receptor can lead to various bone and mineral disorders [5, 6]. About half the total serum Ca^{2+} is bound to protein and the free Ca^{2+} concentration is usually close to 1.4 mM. It has been predicted, however, that local free extracellular Ca^{2+} concentrations could vary over quite a wide range, for example, during a period of Ca^{2+} flux into brain cells [7].

Important extracellular processes that require Ca^{2+} include blood clotting/fibrinolysis (e.g. factors VII, IX, protein C, thrombomodulin, fibrinogen), complement activation (e.g. C1r, C1s), cell surface receptor/ligand interactions (e.g. Notch, LDL receptor), cell adhesion (e.g. selectins, integrins, cadherins), enzyme degradation (e.g. metallo-proteinases, phospholipases, lyases), and connective tissue formation (e.g. fibrillin, osteonectin). Several structures of proteins involved in these processes have been determined (see PDB database). The usual calcium ligands are carboxylate and carbonyl groups and the coordination geometry is somewhat flexible, with 6–8 coordinating ligands being found. In extracellular proteins, calcium seems to play a role both in structure stabilisation and in the mediation of protein-protein interactions [3, 8, 9].

Most extracellular proteins in higher organisms are assembled from a limited number of protein domains or modules [10]. In many cases, the calcium-binding activity of a particular extracellular protein has been assigned to particular modules. Well-documented examples include the Gla domain in vitamin K dependent clotting factors [11], the C-type lectin domain in carbohydrate-binding proteins [12], the EC domain of osteonectin or BM-40 [13], the cadherin module [14], the LDL binding module [15] and the EGF module (see below).

2 The EGF Module and the cbEGF Subset

The epidermal growth factor (EGF) like domain is one of the most common types of module that has been identified in extracellular proteins (see review [16]). It appears to have evolved from a common ancestor by gene duplication, point mutations and exon shuffling [17]. Based on its observed distribution, it has been proposed that it plays a general role in adhesion, coagulation and receptor-ligand interactions. Each EGF module includes ~40–45 amino acids with a signature pattern of six cysteine (Cys) residues which normally disulphide bond in a 1–3, 2–4, 5–6 pattern [10]. A distinct subset of these domains has been identified which contains a consensus sequence associated with calcium binding (cb): D/N-x-D/N-E/Q-x_m-D/N*-x_n-Y/F (where m and n are variables, and * indicates possible β-hydroxylation) [18–20]. As shown schematically in Fig. 1, specific carboxylate- and carboxyamide-containing

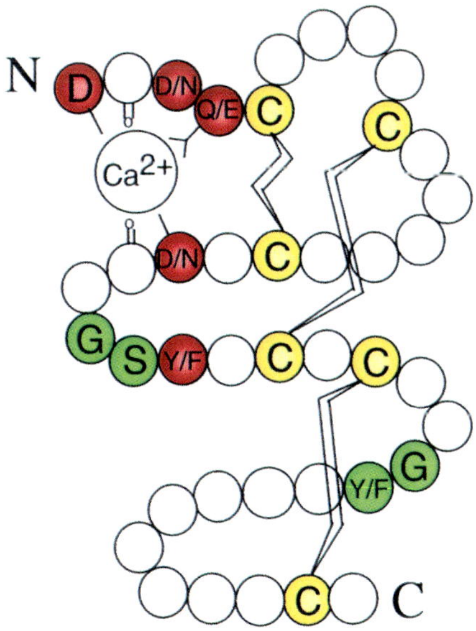

Fig. 1. Schematic illustration of the secondary structure of the calcium-binding EGF-like domain. Each amino acid is shown as a *circle*. Conserved cysteine residues which disulphide bond in a 1–3, 2–4, 5–6 pattern are shaded *yellow*. The calcium-binding consensus sequence, which contains residues important for calcium ligation as well as formation of the calcium-binding site, is shown in *red*. The number of side-chain ligand oxygens donated at a particular position depends on the specific calcium-binding consensus sequence (see text). Other conserved residues, most probably important for folding of the cbEGF domain or stabilising domain-domain interactions, are shown in *green*

side chains are required as calcium ligands, whereas other residues are important for maintaining the geometry of the calcium-binding site (see below). While the specific pattern of carboxyamide and carboxylate side-chain groups may vary from one binding site to the next, a conserved number of side-chain ligand oxygen atoms is required for native calcium affinity [19–22]. Other conserved residues in the cbEGF domain consensus sequence are most probably important for domain folding or domain-domain interactions.

cbEGF domains have been identified in a wide variety of proteins, many of which are involved in blood coagulation, the complement system and the structure and organisation of tissues. A sequence motif defining the cbEGF domain exists in the PROSITE database [23] (accession code: PS01187), which may be used to search protein sequences in the SWISSPROT and TREMBL databases [24] for occurrences of the domain (http://www.expasy.ch/prosite/). The domain organisations of a representative subset of these proteins are shown in Fig. 2. A striking feature of the modular architectures of some cell surface receptor proteins, such as the Notch superfamily, and extracellular matrix proteins, such as the fibrillins and LTBPs, is that cbEGF domains exist as multiple tandem repeats.

cbEGF domain structures from factor VII, factor IX, C1r and fibrillin-1 have been solved by X-ray crystallography and NMR spectroscopy (Table 1). A multiple sequence alignment corresponding to these structures is given in Fig. 3A, and a structural superposition is shown in Fig. 3B. The structures were overlaid with respect to the coordinates of the "EGF core" [25] of the holo N-terminal EGF of factor IX solved to 1.5 Å resolution by X-ray crystallography [26]. The EGF core is a region of the cbEGF that is constant in length in sequences with known structure. It comprises residues between the third Cys and the conserved aromatic residue between the 5–6 disulphide bond, and it includes the major β-hairpin of the cbEGF fold.

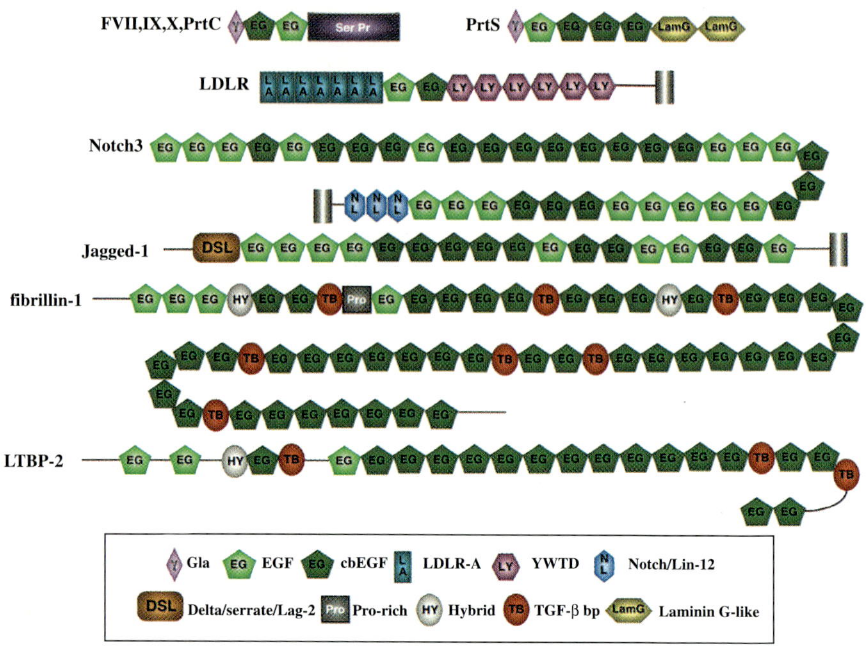

Fig. 2. Modular architectures of a representative set of cbEGF domain containing proteins including vitamin K dependent coagulation factors VII, IX, X and proteins C and S, the low-density lipoprotein receptor (LDLR), the transmembrane cell-signalling molecules Notch3 and Jagged-1, and the extracellular matrix proteins fibrillin-1 and latent transforming growth factor-β binding protein (LTBP)-2. For clarity, only the extracellular portions of the LDLR, Notch3 and Jagged-1 are shown, and the activation domains of FVII, IX, X and protein C (which lie between the EGF and the serine protease domains) are omitted. Domain symbols and abbreviations are according to Bork et al. [10]

Table 1. Structures of calcium-binding EGF-like domains

PDB entry	Description	Method	Apo/holo	Ref
1DAN	Factor VIIa complexed with tissue factor	X-ray (2.0 Å)	Holo	[32]
1BF9	N-terminal cbEGF from human factor VII	NMR	Apo	[31]
1F7M	N-terminal cbEGF from human factor VII	NMR	Holo	[38]
1IXA	N-terminal cbEGF from human factor IX	NMR	Apo	[58]
1PFX	Porcine clotting factor IXa	X-ray (3.0 Å)	Apo	[44]
1EDM	N-terminal cbEGF from human factor IX	X-ray (1.5 Å)	Holo	[26]
1APO	N-terminal cbEGF from human factor X	NMR	Apo	[59]
1CCF	N-terminal cbEGF from human factor X	NMR	Holo	[30]
1WHE	Gla-cbEGF from human factor X	NMR	1Ca[a]	[43]
1APQ	C1r cbEGF	NMR	Apo[b]	[27]
1EMN	cbEGF32–33 from human fibrillin-1	NMR	Holo	[25]

[a] Calcium bound to cbEGF but not to the Gla domain.
[b] For a comparison of the structures of both apo and holo forms, see Ref. 27.

A schematic ribbon drawing of the cbEGF domain topology is shown in Fig. 4. The fold comprises a major and usually a minor double-stranded β-hairpin, and sometimes a short α-helical turn between the first and second Cys residues. The

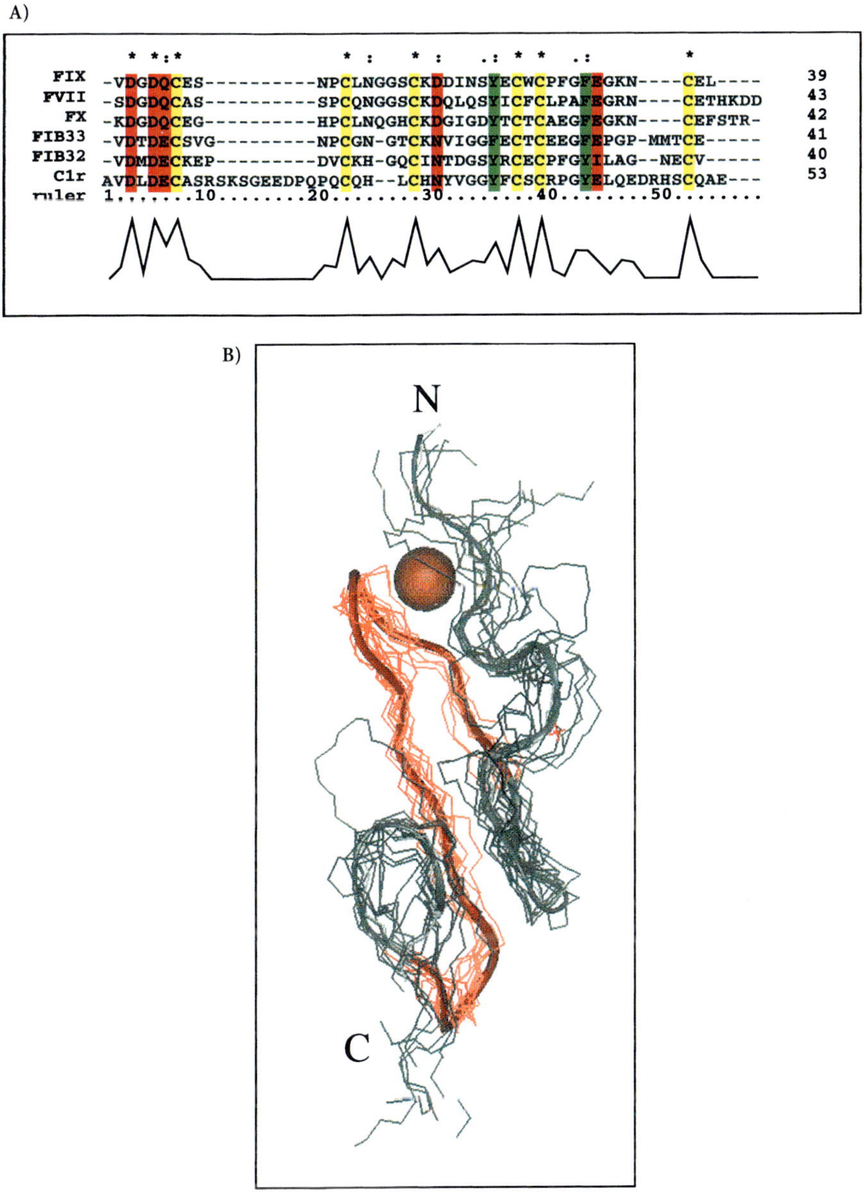

Fig. 3. Multiple alignment of sequences of cbEGF domains with known structures produced using ClustalX (1.8) **A** and superposition of structures **B**. In **A** conserved cysteines, aromatic residues and residues with carboxylate/carboxyamide side chains are shaded *yellow*, *green* and *red*, respectively. *FIX* cbEGF$_N$, factor IX; *FVII* cbEGF$_N$, factor VII; *FX* cbEGF$_N$, factor X; *FIB32* cbEGF 32, fibrillin-1; *FIB33* cbEGF33, fibrillin-1, *C1r* cbEGF, C1r. In **B** the position of the cbEGF core is shown in *red*. The backbone coordinates of FIX, the reference structure for the superposition, are highlighted by a ribbon. Part **B** was created using Insight II (ver. 98, MSI Inc.)

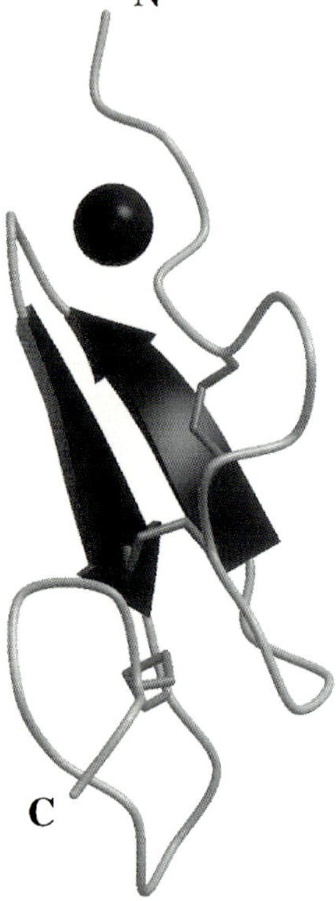

Fig. 4. Schematic ribbon drawing of the cbEGF domain fold, comprising a major double-stranded antiparallel β-sheet and an N-terminal loop which may have α-helical character. The C-terminal region of the domain is sometimes also observed to have β-character (e.g. see Fig. 7C). Consensus disulphide bonds are depicted by *sticks*. The calcium atom bound near the N-terminus is shown as a *sphere*. The structure shown is that of the factor IX $cbEGF_N$ [26]. This figure was rendered [56] from MOLSCRIPT [57] output

coordinates of the 12 structures overlaid in Fig. 3B agree well in the core region, with backbone RMSDs ranging from 0.22–1.6 Å. However, the sequences and structures are more variant in the flanking regions, which may be relevant to the specificity of the cbEGF domain function in different proteins. For example, the C1r cbEGF domain contains an unusually large loop between the first and second Cys residues, which does not possess a unique structure. Consequently, this loop has been postulated to play a role in domain-domain interactions within C1r or protein-protein interactions within the C1 complement complex [27]. Interestingly, the conformation of this loop does not appear to be stabilised by calcium binding.

cbEGF domain sequences contain fewer hydrophobic residues than would be expected based on the size of the module, which gives a domain lacking an appreciable hydrophobic core [26, 28]. The structure is principally stabilised by the three consensus disulphide bonds and by calcium binding to the amino terminus. Even in the highest resolution X-ray crystal structure of a cbEGF domain that has been determined, almost one quarter of the residues lie in less favoured regions of the Ramachandran plot, which is well below what would normally be expected for a well-folded globular protein [29].

3 Calcium Binding to EGFs

Comparative analysis of structures of the same cbEGF domain in the apo and holo states has demonstrated that only minimal conformational changes are associated with calcium binding by the isolated domain [27, 30, 31]. Specifically, residues in the vicinity of the calcium ligands move closer together, and the structural definition of the N-terminus improves. The specific ligand geometry for calcium binding has been elucidated by crystal structures of cbEGF domains from factors VII and IX [26, 32]. Calcium is coordinated by a pentagonal bipyramidal arrangement of oxygen atoms with an average Ca^{2+}—O distance of 2.3 Å. As shown in Fig. 5, the ligands typically comprise four side-chain oxygens, two backbone oxygens and one oxygen in a water molecule. However, packing of two domains in the crystal asymmetric unit of the cbEGF domain from factor IX leads to the replacement of the water molecule ligand by a side-chain oxygen from the adjacent domain (see below and Fig. 8). These observations have led to suggestions that calcium can directly mediate intramolecular domain-domain interactions [26] and that displacement of water molecule ligands from EGF calcium-binding sites may play a general role in protein-protein interactions [25, 30]. Interestingly, the second D/N residue of the calcium-binding consensus sequence (D/N-x-D/N-E/Q-x_m-D/N*-x_n-Y/F, see above) plays an indirect role in calcium binding: the carboxylate side chain forms an ASX turn with the backbone amide nitrogen of the i + 2 Cys residue. The conserved aromatic residue also plays a structural role in maintaining the geometry of the calcium-binding site via hydrophobic packing interactions [26].

Calcium binding by an EGF domain was first demonstrated in protein C using a calcium-sensitive electrode [33]. Since then, a number of calcium dissociation constants for EGF domains have been measured using one- (1-D) and two-dimensional (2-D) NMR, intrinsic protein fluorescence, equilibrium dialysis and chro-

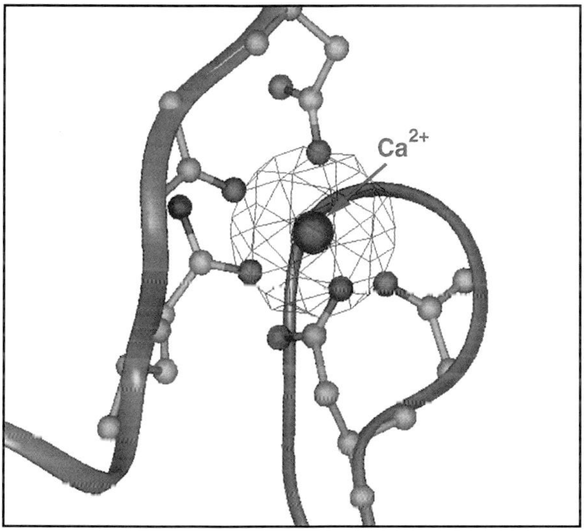

Fig. 5. Calcium coordination by a cbEGF domain. Ligands typically include four side-chain oxygens, two backbone oxygens and one oxygen from a water molecule (not shown) organised in a pentagonal bipyramidal arrangement. The specific number of side-chain ligand oxygens provided at a particular position varies as a function of the calcium-binding sequence. The calcium-binding site shown is from factor IX cbEGF$_N$ [26]. This figure was created using Insight II (ver. 98, MSI Inc.)

mophoric chelator methods (summarised in Table 2). Both the accuracy and range of possible measurements depend on the sensitivity of the method used. K_d measurement using NMR is based on changes of crosspeak position (usually that corresponding to the ring protons of the conserved aromatic residue in the cbEGF consensus sequence) as a function of added calcium. Since signals in the NMR spectrum may be assigned sequence specifically, NMR is the only method which allows definitive assignment of K_d values to individual domains when studying constructs containing multiple binding sites. NMR is, however, a relatively insensitive method and may only be used reliably to measure low affinities, corresponding to K_d values $\geq\sim 100$ μM, since the protein sample concentration is >100 μM. More dilute samples may be used in intrinsic protein fluoresence studies, but since the cbEGF domain lacks an appreciable hydrophobic core, this technique is usually only useful for measurement of slightly higher cbEGF affinities than NMR ($K_d \geq \sim 1$ μM).

Equilibrium dialysis is less informative than NMR since it produces an average dissociation constant when multiple sites are present, but requires less specialised equipment and lower protein concentrations. Since chromophoric chelators exist with a wide range of calcium affinities, measurement of changes in the absorbance properties of these compounds, which bind calcium in competition with the protein of interest, may be used to probe dissociation constants in the range ~1 nM–

Table 2. Calcium dissociation constants measured for cbEGF domains[a]

Construct	K_d (s)	pH	I	Method	Ref
Protein C $cbEGF_N$	100 μM	7.5	0.1	Calcium-sensitive electrode	[33]
FX $cbEGF_N$	2 mM	7.4	0.15	1-D NMR	[60]
FX Gla-$cbEGF^b_N$	120 μM	7.5	0.1	Fluorescence	[61]
FIX $cbEGF_N$	1.8 μM	7.4	0.15	1-D NMR	[62]
HsFIXcbEGF	1 mM	7.4	0.15	Fluorescence	[63]
Notch cbEGF11–12	1:34 μM; 2: > 0.25 mM	7.5	0.15	Chromophoric chelator	[34]
Notch cbEGF10–13	1:3.1 μM; 2:0.16 mM; 3: > 0.25 mM	7.5	0.15	Chromophoric chelator	[34]
C1r cbEGF	10 mM	6.6	0	1-D NMR	[64]
C1r/C1s CUB-cbEGF	32–38 μM	7.4	0.15	Equilibrium dialysis	[65]
Protein S cbEGF3	6 mM	7.5	0.15	1-D NMR	[66]
Protein S cbEGF4	~8 mM	7.5	0.15	1-D NMR	[66]
Protein S cbEGF1–4	1:20 nM; 2:2 μM; 3:5 μM	7.5	0.15	Chromophoric chelator	[37]
Protein S cbEGF2–3	1:15 μM; 2: ≥ 10 μM	7.5	0.15	Chromophoric chelator	[37]
Protein S cbEGF2–4	1:1 μM; 2: ≥ 10 μM; 3: ≥ 10 μM	7.5	0.15	Chromophoric chelator	[37]
Fibrillin cbEGF12–13	12:1.6 mM; 13: ≤ 30 μM	6.5	0.15	2-D NMR/ fluoresence	[36]
Fibrillin cbEGF13–14	13:3 mM; 14: ≤ 100 μM	6.5	0.15	2-D NMR	[67]
Fibrillin cbEGF22–23	~400 μM (n = 2)	7.5	0.15	Equilibrium dialysis	[68]
Fibrillin cbEGF22–31	~500 μM (n = 10)	7.5	0.15	Equilibrium dialysis	[68]
Fibrillin cbEGF25–31	250 μM (n = 6–7)	7.5	0.17	Equilibrium dialysis	[41]
Fibrillin cbEGF32	4.3 mM	7.4	0.15	1-D NMR	[21]
Fibrillin TB6-cbEGF32	1.6 mM	6.5	0.15	2-D NMR	[22]
Fibrillin cbEGF32–33	32:9.2 mM; 33:350 μM	6.5	0.15	2-D NMR	[35]

[a] In studies where calcium dissociation constants were measured under varying conditions of pH and ionic strength, the value corresponding to the most physiological conditions is reported.
[b] In this construct the Gla residues were decarboxylated to Glu so the binding to the cbEGF domain could be monitored exclusively.

250 μM. However, the measurement of high affinities is technically demanding, and relies on preparation of completely calcium-free protein and buffer solutions, as well as very accurate concentrations of calcium stocks.

Calcium dissociation constants measured for N-terminal or isolated cbEGF domains are in the range 1–10 mM K_d. Differences of up to one order of magnitude for the same cbEGF can be explained on the basis of different sample conditions [19]. It is likely that for different cbEGFs calcium affinities are sequence dependent. Attempts have been made to rationalise affinities in terms of charge distribution and flexibility of the site. Selander-Sunnerhagen et al. [30] noted that hydration should be repulsive to the calcium, since the total charge of the calcium-binding region is negative even with a buried calcium. In such a case, addition of an adjacent hydrophobic group, particularly at the D/N-x-D/N position of the calcium-binding consensus sequence, should enhance the affinity of the binding site. Rand et al. [34] suggested that differences in electronegative potentials for cbEGF domain pairs might provide a basis for interpretation of differing calcium affinities. Finally, energetically unfavourable changes in entropy associated with ordering of a flexible calcium-binding site have been postulated, by Downing et al. [25], to result in higher calcium dissociation constants based on K_d values [35] and structural comparison of the two sites in a cbEGF domain pair from fibrillin-1.

Comparison of the calcium dissociation constants for the same cbEGF domain with a free N-terminus versus one that is N-terminally linked to an adjacent domain, in constructs from factor IX, Notch, C1r, protein S and fibrillin-1, demonstrates that N-terminal domain linkage is always accompanied by an increase in calcium-binding affinity. However, as highlighted by Smallridge et al. [36], the magnitude of the affinity enhancement appears to depend on the extent of pairwise domain interactions, and the flexibility of the domain-cbEGF linkage. These observations lend support to the idea that increased entropic energetic penalties correlate with higher calcium dissociation constants. Although no structure of a covalently linked domain pair with interdomain calcium ligation has been reported to date (as suggested by the crystal packing of factor IX cbEGF domain, see below), the possibility exists that this sort of interaction may also increase cbEGF domain affinity. Based on studies of cbEGF domain pairs and triplets from protein S, it was proposed that both N to C and C to N terminal stabilisation of calcium-binding sites occurs [37]. This was not observed to be the case for fibrillin-1 cbEGF domain pairs, but triplet constructs have not yet been studied in this protein [36]. Since the structure of protein S is not known, these observations may reflect differences in the structures of the two proteins.

4
Post-Translational Modifications of cbEGFs

A number of post-translational modifications, which include O-fucosylation, O-glucosylation and Asn/Asp β-hydroxylation, may occur in cbEGF domains, although the extent to which a domain may be modified is variable. Kao et al. [38] investigated the consequences of O-fucosylation on the first EGF-like domain from factor VII; they found that, while the overall structures of the two forms of the domain are very similar even in the region of the fucosylation site, the calcium affinity of the domain is enhanced two-fold by modification of Ser-60. The fucosylation site was observed to occur in the context of a type I′ β-turn, and it was suggested that this β-turn structure is essential for recognition of the site by O-fucosyltransferase. The modification of the calcium-binding properties of the cbEGF by fucosylation may be important in determining the affinity of factor VII for tissue factor and the amidolytic activity of the protease domain (see below).

β-Hydroxylation of Asp/Asn is associated with the consensus sequence C-x-D/N-x_4-Y/F-x-C-x-C. This motif has only been identified in cbEGF domain sequences; however, these sites are sometimes only partially hydroxylated, or even not at all [39–41]. Initially it was proposed that the β-hydroxyl group could act as a calcium ligand. However, neither the calcium-binding properties nor the tertiary structure of cbEGF domains from factors IX and X were observed to change significantly in the presence of this post-translational modification [42]. High-resolution structural studies have shown that the β-hydroxyl group points away from the calcium-binding site [26, 30], and hence may play a role in tertiary interactions such as hydrogen bonding or intermolecular calcium ligation [25].

5 Mutations to cbEGF Domains in Human Disease

Amino acid changes in cbEGF domains have been linked to numerous inherited diseases affecting blood clotting, connective tissue, cardiovascular and nervous systems. The affected proteins and disorders are summarised in Table 3. Mutations affecting cbEGFs in the vitamin K dependent proteases, the LDLR, Notch3, Jagged-1, fibrillins-1 and -2 and LTBP-2 may be classified into three groups based on their predicted consequences. Mutations which either modify a conserved Cys engaged in a disulphide bond, or introduce a Cys into a noncanonical position are probably associated with domain misfolding. Amino acid changes in the calcium-binding consensus sequence are likely to result in reduced calcium affinity and greater interdomain linker flexibility [19, 21, 22]. The consequences of the third group of

Table 3. Heritable genetic disorders associated with mutations to cbEGF domains

Protein(s)	Disease(s)	OMIM #[a]	Ref
Factor VII	Factor VII deficiency	227500	[69]
Factor IX	Haemophilia B	306900	[70]
Protein C	Protein C deficiency	176860	[71]
Protein S	Protein S deficiency	176880	[72]
LDL receptor	Familial hypercholesterolaemia	143890	[73][b]
Notch	CADASIL	125310/600276	[74]
Jagged-1	Alagille syndrome	118450/601920	[75]
Fibrillin-1 and 2, LTBP-2	Marfan syndrome and related phenotypes	154700/121050	[76–78]

[a] OMIM: Online Mendelian inheritance in man database, available: http://www3.ncbi.nlm.nih.gov:80/Omim.
[b] See also: http://www.ucl.ac.uk/fh/genebook.html.

mutations are not so easily predicted, and these need to be considered on a case-by-case basis.

Based on the solution structure of a pair of cbEGF domains from human fibrillin-1 (see below), it was predicted that specific mutations at the end of the major β-sheet will disrupt pairwise cbEGF domain interactions (fibrillin-1: I1175T, P1424A, G1910V, G2627R; fibrillin-2: I1092T). In addition the structure of the soluble tissue factor-factor VIIa complex (see also below and Fig. 6) demonstrated that R79 (which may be mutated in factor VII deficiency) is engaged in a salt bridge with Glu24 of soluble tissue factor. So, while a few of the mutations in the third group have been explained by structural studies, several others, whose consequences are yet unknown, still remain.

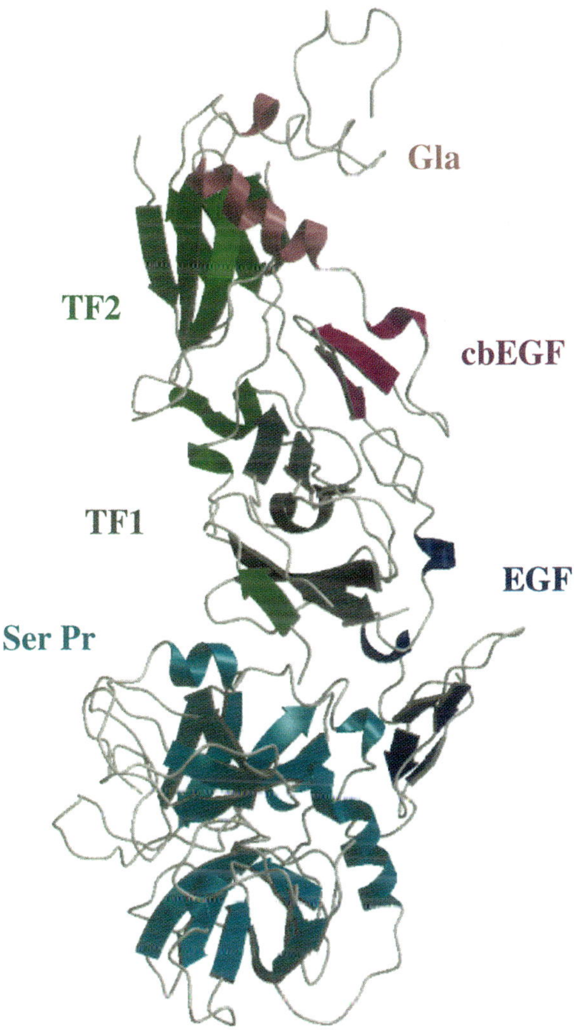

Fig. 6. Schematic ribbon drawing of the soluble tissue factor-factor VIIa complex. The two fibronectin type III domains of sTF are coloured in *dark* and *light green*, and are labelled TF1 and TF2. The back face of the major β-sheet of the factor VIIa cbEGF domain packs against the interface of TF1 and TF2. Only fragments of the Gla domain are shown because in the structure determination insufficient electron density was observed to trace the entire backbone [32]. This figure was rendered [56] from MOLSCRIPT [57] output

6
Assembly of cbEGF Domains

Most structural data relevant to the structures of cbEGF domains in intact molecules and their roles in higher order complexes have been obtained via studies of proteins involved in the blood coagulation cascade. The vitamin K dependent coagulation factors (VII, IX and X and protein C) have identical domain organisations, comprising an N-terminal γ-carboxyglutamic acid (Gla) domain followed by a cbEGF-EGF pair, an activation domain and a C-terminal serine protease (SP) domain (see Fig. 2). All of these blood coagulation factors circulate as single-chain zymogens; they are activated via cleavage events, including lysis of a bond in the activation domain to produce light and heavy chains. Calcium binding to the EGF domain in these proteins is critical for correct function, as evidenced by the identification of disease-causing mutations which affect calcium ligands (see above). The precise role of the calcium-EGF interaction in these proteins was first elucidated by Sunnerhagen et al. [43], who showed that binding results in a more fixed orientation of the two domains, which presumably reduces the entropic energetic penalty in formation of a higher order complex.

The only known structure of a complex including a cbEGF domain is of soluble tissue factor (sTF) and factor VIIa [32]. Factor VII is activated to factor VIIa upon TF exposure due to vessel injury or rupture of atherosclerotic plaques. Formation of the TF-factorVIIa complex, which is only active in the presence of calcium, initiates the blood coagulation process. Soluble TF (sTF, the extracellular region of tissue factor) contains two fibronectin type-III domains, which adopt a well-defined orientation that is unchanged upon complex formation. In contrast, the flexible factor VIIa structure is stabilised by interaction with sTF, mainly via specific contacts with the cbEGF domain which bury 823 Å of surface area. As shown in Fig. 6, the back face of the major double-stranded β-sheet of the cbEGF domain packs against the interface of the two fibronectin type III modules of sTF. This face of the cbEGF domain is not particularly hydrophobic in character, and the binding involves a mixture of polar and non-polar interactions. Consequently it would have been difficult to identify this surface patch of the cbEGF as involved in complex formation based on its surface properties. Since this is the only known structure of a cbEGF in complex, it will be interesting to see in the future whether or not binding generally involves this face of the domain.

As expected, both O-glycosylated sites on the cbEGF domain are solvent exposed in the complex. Hence O-glycosylation does not appear to play an important structural role in the sTF-factor VIIa complex. However, based on calcium-binding studies, it has been proposed that fucosylation may affect the interaction of these two molecules by influencing cbEGF calcium affinity [38]. Since calcium binding would be expected to stabilise the N-terminal region of the major β-hairpin, increased calcium affinity may decrease the entropic penalty associated with complex formation and thereby promote amidolytic activity.

In the sTF-factorVIIa complex, there is no interface between the two EGF domains. In contrast, in the X-ray structure of apo clotting factor IXa, the two domains in the cbEGF-EGF pair are arranged at an angle of 110°, stabilised by a conserved and factor IX specific interdomain salt bridge [44] (Fig. 7A). Whether or not the salt-

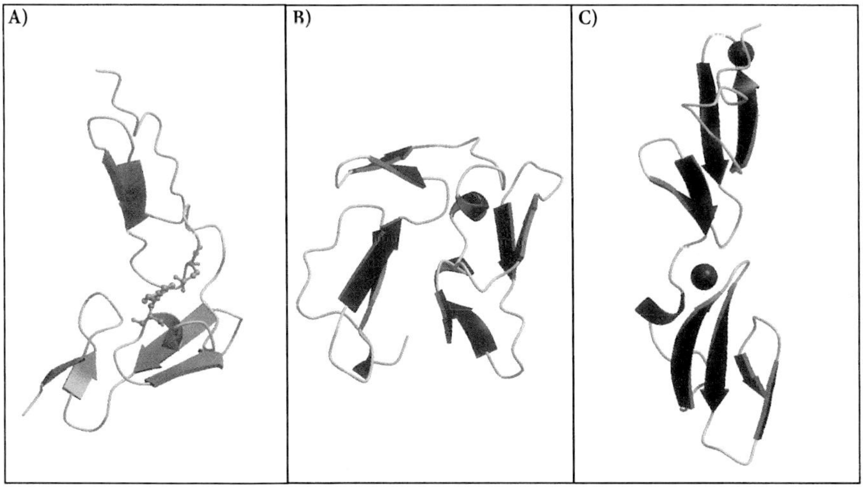

Fig. 7. Schematic ribbon drawings of tandem EGF pairs from **A** factor IXa [44], **B** merozoite surface plasmodium-1 [46], and **C** fibrillin-1 [25]. Residues involved in the salt-bridging interaction which stabilises the cbEGF-EGF orientation in (**A**) are shown in ball-and-stick. Calcium atoms are rendered as *spheres* in (**C**). Comparison of the three pairs, cbEGF-EGF (**A**), EGF-EGF (**B**) and cbEGF-cbEGF (**C**), provides insight into the role of calcium in stabilising the N-terminal region of the cbEGF domain in an extended conformation. This figure was rendered [56] from MOLSCRIPT [57] output

bridging interaction significantly restricts the flexiblity of the interdomain linkage of this pair in solution has not been resolved. This is the only structure of a cbEGF-EGF pair which has been reported to date, although recently the structure of a non-calcium-binding EGF pair from *Plasmodium falciparum* merozoite surface protein-1 was solved by NMR and X-ray crystallography [45, 46]. The two domains adopt a U-shaped fold, with a large area of hydrophobic core buried in the domain interface, as shown in Fig. 7B.

Comparison of this structure with that of a tandem pair of cbEGF domains from fibrillin-1 sheds light on the role of calcium in orienting the N-terminus of the domain. The two cbEGF domains are oriented in a rigid, rod-like conformation as illustrated in Fig. 7C, stabilised by hydrophobic packing interactions and calcium binding by the C-terminal domain [25]. Based on sequence analysis, it was predicted that all tandem cbEGF domain regions in the fibrillins and several other proteins would have the same organisation. Calcium-dependent changes in the fibrillin-1 ultrastructure in microfibrils and fragments [47–49] support this conclusion. Thus this structure has provided significant insight into the structure of the intact fibrillin molecules as well as the structural consequences associated with disease-causing mutations to these proteins. In addition, a recent structure determination of an EGF-cbEGF domain pair from the LDL receptor has revealed a very similar domain organisation, supporting the theory that pairwise domain organisation can be predicted at the sequence level for functionally distinct classes of proteins [25, 50].

Calcium-dependent interactions between members of the Notch family of signalling proteins play a major role in specification of cell fate (reviewed in [51, 52]).

This is another class of proteins, like the fibrillins, which is characterised by multiple tandem repeats of cbEGF domains. However, sequence data suggest that the mode of pairwise domain linkage in these proteins may differ from that observed for fibrillin-1 [25]. These domain pairs contain an additional residue in the interdomain linkage, and a pattern of amino acid conservation more similar to cbEGF domains from the vitamin K dependent blood coagulation factors. In the crystal structure of the N-terminal cbEGF domain from factor IX, it was observed that the two domains in the crystal asymmetric unit adopt a V-shaped arrangement, with the N- and C-termini in close proximity, and a calcium ligand donated by an Asn residue from the adjacent domain (Fig. 8). Intriguingly, the Asn involved in this interaction is conserved in cbEGF domain pair sequences from the Notch family of proteins. Since the region involved in Notch-ligand interactions has been narrowed to a single pair of cbEGF domains [53], it will be of particular interest to see whether or not the conformation of this pair resembles the crystal packing of the factor IX cbEGF domain.

C1r and C1s are two cbEGF-containing proteins which, with C1q, form the human C1 complement complex (see review [54]). The C1 complex triggers the classical pathway of complement in response to infection and to the formation of antibody-antigen complexes. C1r and C1s each contain an N-terminal CUB-cbEGF-CUB module triplet known as the α-region. These two serine proteases are only catalytically active in the context of the C1s-C1r-C1r-C1s complex, which forms via calcium-dependent association of the α-regions. Recent studies of C1r fragments comprising the N-terminal CUB module, the cbEGF and the CUB-cbEGF pair have probed the role of pairwise domain interactions in the association of C1r and C1s [55]. These experiments have shown that the CUB-cbEGF pair adopts a more compact conformation in the presence of calcium, and that the CUB-cbEGF pair is the minimal segment for Ca^{2+}-dependent interaction with C1s.

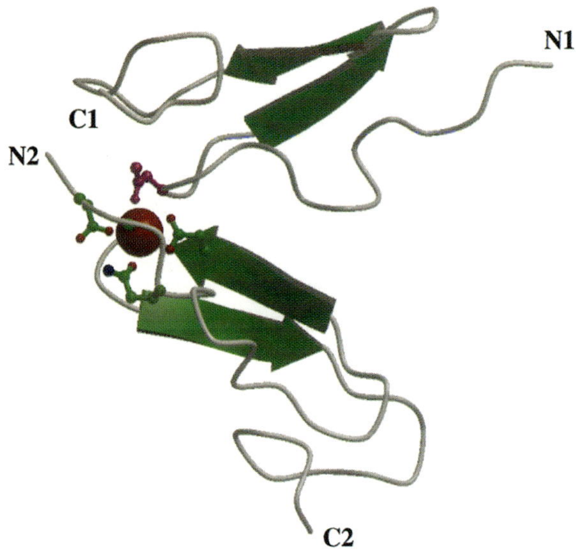

Fig. 8. Crystal packing of the factor IX isolated cbEGF$_N$. Two copies of the domain were observed in the crystal asymmetric unit to pack with their N- and C-termini in close spatial proximity, and with an Asn residue donated as a calcium ligand from one domain to the calcium-binding site of the next (shown in magenta) [26]. The V-shaped arrangement of the two domains has been proposed as a model for tandem cbEGF domains from the Notch superfamily and related proteins [25, 26]. This figure was rendered [56] from MOLSCRIPT [57] output

7 Conclusions

EGF domains are one of the most abundant extracellular protein modules. A particular type of EGF which has evolved to bind calcium is found in a variety of proteins involved in diverse biological processes. The identification of missense mutations within cbEGF domains which cause numerous genetic disorders has led to extensive biophysical characterisation of their properties in terms of structure, post-translational modifications, calcium binding, and protein-protein interactions. A review of the body of research associated with cbEGF domains has revealed some general features of the function of the calcium-binding motif. Calcium binding by the EGF domain appears to play a general role in restricting the conformational flexibility of interdomain linkages and/or orienting domain pairs. Modulation of domain-domain linker properties may play a role in determining the energetics of tertiary protein interactions and the biomechanical properties of proteins. The stability of the cbEGF domain N-terminus appears to be related to the affinity with which the domain binds. Furthermore, domain-domain packing interactions appear to enhance the calcium-binding properties of C-terminally linked cbEGF domains, at least in part via stabilisation of the calcium-binding site.

Acknowledgements. This is a contribution from the Oxford Centre for Molecular Sciences that is supported by BBSRC, MRC and EPSRC. We also thank the Welcome Trust, BBSRC, MRC and the British Heart Foundation for separate support. AKD is a Wellcome Trust Senior Research Fellow.

References

1. Bootman MD, Berridge MJ (1995) Cell 83:675
2. Kretsinger RH (1996) Nat Struct Biol 3:12
3. Evenäs J, Malmendal A, Forsén S (1998) Curr Opin Chem Biol 2:293
4. Williams RJP (1998) Biochim Biophys Acta 1448:153
5. Brown EM, Vassilev PM, Hebert SC (1995) Cell 83:679
6. Cole DE, Peltkova VD, Rubin LA, Hawker GA, Veith R, Liew CC, Hwang DM, Evorski J, Hendy GN (1999) Lancet 353:112
7. Egelman DM, Montague PR (1999) Biophys J 76:1856
8. Beckingham K, Quing Lu A, Andruss AF (1998) Biometals 11:359
9. Maurer P, Hohenester E (1997) Matrix Biol 15:569
10. Bork P, Downing AK, Kieffer B, Campbell ID (1996) Q Rev Biophys 29:119
11. Stenflo J (1999) Crit Rev Eukaryot Gene Exp 9:59
12. Weiss WI, Taylor ME, Drickamer K (1998) Immunol Rev 163:19
13. Hohenester E, Maurer P, Hohenadl C, Timpl R, Jansonius JN, Engel J (1996) Nat Struct Biol 3:67
14. Alattia JR, Kurokawa H, Ikura M (1999) Cell Mol Life Sci 55:359
15. Fass D, Blacklow S, Kim PS, Berger JM (1997) Nature 388:691
16. Campbell ID, Bork P (1993) Curr Opin Struct Biol 3:385
17. Patthy L (1991) Curr Opin Struct Biol 1:351
18. Rees DJG, Jones IM, Handford PA, Walter SJ, Esnouf MP, Smith KJ, Brownlee GG (1988) EMBO J 7:2053
19. Handford PA, Mayhew M, Baron M, Winship PR, Campbell ID, Brownlee GG (1991) Nature 351:164

20. Mayhew M, Handford P, Baron M, Tse AGD, Campbell ID, Brownlee GG (1992) Protein Eng 5:489
21. Handford PA, Downing AK, Rao Z, Hewett DR, Sykes BC, Kielty CM (1995) J Biol Chem 270:6751
22. Kettle S, Yuan X, Grundy G, Knott V, Downing AK, Handford PA (1999) J Mol Biol 285:1277
23. Hofmann K, Bucher P, Falquet L, Bairoch A (1999) Nucleic Acids Res 27:215
24. Bairoch A, Apweiler R (1999) Nucleic Acids Res 27:49
25. Downing AK, Knott V, Werner JM, Cardy CM, Campbell ID, Handford PA (1996) Cell 85:597
26. Rao Z, Handford P, Mayhew M, Knott V, Brownlee GG, Stuart D (1995) Cell 82:131
27. Bersch B, Hernandez J-F, Marion D, Arlaud GJ (1998) Biochemistry 37:1204
28. McCaldon P, Argos P (1988) Proteins 4:99
29. Laskowski RA, Macarthur MW, Moss DS, Thornton JM (1993) J Appl Crystallogr 26:283
30. Selander-Sunnerhagen M, Ullner M, Persson E, Teleman O, Stenflo J, Drakenberg T (1992) J Biol Chem 267:19642
31. Muranyi A, Finn BE, Gippert GP, Forsen S, Stenflo J, Drakenberg T (1998) Biochemistry 37:10605
32. Banner DW, D'Arcy A, Chene C, Winkler FK, Guha A, Konigsberg WH, Nemerson Y, Kirchhofer D (1996) Nature 380:41
33. Ohlin AK, Linse S, Stenflo J (1988) J Biol Chem 263:7411
34. Rand MD, Lindblom A, Carlson J, Villoutreix BO, Stenflo J (1997) Protein Sci 6:2059
35. Knott V, Downing AK, Cardy CM, Handford P (1996) J Mol Biol 255:22
36. Smallridge RS, Whiteman P, Doering K, Handford PA, Downing AK (1999) J Mol Biol 286:661
37. Stenberg Y, Linse S, Drakenberg T, Stenflo J (1997) J Biol Chem 272:23255
38. Kao Y-H, Lee GF, Wang Y, Starovasnik MA, Kelley RF, Spellman MW, Lerner L (1999) Biochemistry 38:7097
39. Thim L, Bjoern S, Christensen M, Nicolaisen EM, Lundhansen T, Pedersen AH, Hedner U (1988) Biochemistry 27:7785
40. Stenflo J, Holme E, Lindstedt S, Chandramouli N, Huang LHT, Tam JP, Merrifield RB (1989) Proc Natl Acad Sci USA 86:444
41. Glanville RW, Qian R-Q, McClure DW, Maslen CL (1994) J Biol Chem 269:26630
42. Selander-Sunnerhagen M, Persson E, Dahlqvist I, Drakenberg T, Stenflo J, Mayhew M, Robin M, Handford P, Tilley JW, Campbell ID, Brownlee GG (1993) J Biol Chem 268:23339
43. Sunnerhagen M, Olah GA, Stenflo J, Forsen S, Drakenberg T, Trewhella J (1996) Biochemistry 35:11547
44. Brandstetter H, Bauer M, Huber R, Lollar P, Bode W (1995) Proc Natl Acad Sci USA 92:9796
45. Morgan WD, Birdsall B, Frenkiel TA, Gradwell MG, Burghaus PA, Syed SEH, Uthaipibull C, Holder AA, Feeney J (1999) J Mol Biol 289:113
46. Chittara V, Holm I, Bentley GA, Petres S, Longacre S (1999) Mol Cell 3:457
47. Kielty CM, Shuttleworth CA (1993) FEBS Lett 336:323
48. Reinhardt DP, Mechling DE, Boswell BA, Keene DR, Sakai LY, Bachinger HP (1997) J Biol Chem 272:7368
49. Cardy CM, Handford PA (1998) J Mol Biol 276:855
50. Downing AK (1999) University of Oxford, personal communication
51. Gridley T (1997) Mol Cell Neurosci 9:103
52. Weinmaster G (1997) Mol Cell Neurosci 9:91
53. Rebay I, Fleming RJ, Fehon RG, Cherbas L, Cherbas P, Artavanis-Tsakonas S (1991) Cell 67:687
54. Arlaud GJ, Colomb MG, Gagnon J (1987) Immunol Today 8:106
55. Thielens NM, Enrie K, Lacroix M, Jaquinod M, Hernandez J-F, Esser AF, Arlaud GJ (1999) J Biol Chem 274:9149
56. Merritt EA, Murphy MEP (1994) Acta Crystallogr D 50:869
57. Kraulis PJ (1991) J Appl Crystallogr 24:946
58. Baron M, Norman DG, Harvey TS, Handford PA, Mayhew M, Tse AGD, Brownlee GG, Campbell ID (1992) Protein Sci 1:81
59. Ullner M, Selander M, Persson E, Stenflo J, Drakkenberg T, Teleman O (1992) Biochemistry 31:5974
60. Persson E, Selander M, Linse S, Drakenberg T, Ohlin AK, Stenflo J (1989) J Biol Chem 264: 16897

61. Valcarce C, Selander-Sunnerhagen M, Tamlitz AM, Drakenberg T, Bjork I, Stenflo J (1993) J Biol Chem 268:26673
62. Handford PA, Baron M, Mayhew M, Willis A, Beesley T, Brownlee GG, Campbell ID (1990) EMBO J 9:475
63. Hughes PE, Morgan G, Rooney EK, Brownlee GG, Handford P (1993) J Biol Chem 268:17727
64. Hernandez JF, Bersch B, Petillot Y, Gagnon J, Arlaud GJ (1997) J Pept Res 49:221
65. Thielens NM, Aude CA, Lacroix MB, Gagnon J, Arlaud GJ (1990) J Biol Chem 265:14469
66. Stenberg Y, Julenius K, Dahlqvist I, Drakenberg T, Stenflo J (1997) Eur J Biochem 248:163
67. Whiteman P, Downing AK, Smallridge R, Winship PR, Handford PA (1998) J Biol Chem 273:7807
68. Reinhardt DP, Keene DR, Corson GM, Poschl E, Bachinger HP, Gambee JE, Sakai LY (1996) J Mol Biol 258:104
69. Tuddenham EGD, Pemberton S, Cooper DN (1995) Thromb Haemost 74:313
70. Gianelli F, Green PM, Sommer SS, Poon M-C, Ludwig M, Schwaab R, Reitsma PH, Goossens M, Yoshioka A, Figueiredo MS, Brownlee GG (1998) Nucleic Acids Res 26:265
71. Reitsma PH, Bernardi F, Doig RG, Gandrille S, Greengard JS, Ireland H, Krawczak M, Lind B, Long GL, Poort SR, Saito H, Sala N, Witt I, Cooper DN (1995) Thromb Haemost 73:876
72. Gandrille S, Borgel D, Ireland H, Lane DA, Simmonds R, Reitsma PH, Mannhalter C, Pabinger I, Saito H, Suzuki K, Formstone C, Cooper DN, Espinosa Y, Sala N, Bernardi F, Alach M (1997) Thromb Haemost 77:1201
73. Varret M, Rabes J-P, Thiart R, Kotze MJ, Baron H, Cenarro A, Descamps O, Ebhardt M, Hondelijn J-C, Kostner GM, Miyake Y, Pocovi M, Schmidt H, Schmidt H, Schuster H, Stuhrmann M, Yamamura T, Junien C, Beroud C, Boileau C (1998) Nucleic Acids Res 26:248
74. Joutel A, Vahedi K, Corpechot C, Troesch A, Chabriat H, Vayssiere C, Cruaud C, Maciazek J, Weissenbach J, Bousser MG, Bach JF, Tournier Lasserve E (1997) Lancet 350:1511
75. Crosnier C, Driancourt C, Raynaud N, Dhorne-Pollet S, Pollet N, Bernard O, Hadchouel M, Meunier-Rotival M (1999) Gastroenterology 116:1141
76. Collod-Beroud G, Beroud C, Ades L, Black C, Boxer M, Brocks DJH, Holman KJ, dePaepe A, Francke U, Grau U, Hayward C, Klein HG, Liu WG, Nuytinck L, Peltonen L, Perez ABA, Rantamaki T, Junien C, Boileau C (1998) Nucleic Acids Res 26:229
77. Park E-S, Putnam EA, Chitayat D, Child A, Milewicz DM (1998) Am J Med Genet 78:350
78. Matthews KR, Godfrey M (1997) Am J Hum Genet 61:228

Calmodulin-Dependent Protein Kinases

Joachim Krebs

Institute of Biochemistry, Swiss Federal Institute of Technology (ETH), Universitaetsstr. 16, CH-8092 Zurich, Switzerland (*E-mail: krebs@bc.biol.ethz.ch*)

Protein-dependent kinases are key regulators of many cellular events. One important class, the calmodulin-dependent protein kinases (CaMkinases), is dependent on changes of the intracellular concentrations of the second messenger Ca^{2+}. In this chapter the structural and functional properties of CaMkinases are summarized. The different kinases known to date can be either monospecific for one substrate or multifunctional, recognizing a number of different substrates. CaMkinases are involved in the regulation of many different cellular processes such as neurotransmitter release, muscle contraction, cell proliferation, gene expression, and long term potentiation and depression. Their participation in these different regulatory pathways are evaluated.

Keywords. Calmodulin, Calmodulin-dependent protein kinase (CaMK), DAP-kinase, Gene expression, Synaptic plasticity, Apoptosis

1 Introduction

Numerous complex biological processes such as proliferation, differentiation, and apoptosis are controlled by a small number of second messengers. Ca^{2+}, as one of the most versatile second messengers, plays its pivotal role through a specific class of proteins, the so-called EF-hand type Ca^{2+}-binding proteins [1]. The most prominent representative of the intracellular Ca^{2+}-binding proteins is calmodulin (CaM) [2] which binds to and activates a great variety of enzymes, notably a class of different protein kinases, i.e., CaM-dependent protein kinases (CaMK), and the phosphatase calcineurin [3–5]. Protein kinases catalyze the transfer of a phosphoryl group from ATP to the appropriate residue of a protein, thereby regulating the biological activity by a stoichiometric covalent modification of the protein. Since protein kinases and phosphatases are key modulators of a variety of cellular processes, it is not surprising that CaMkinases are involved in the regulation of many different cellular processes such as neurotransmitter release, muscle contraction, cell proliferation, and gene expression. A number of different CaMkinases are known to date, all belonging to the class of enzymes phosphorylating serine-threonine residues. The targets can be either of multiple sources (e.g., CaMK I (V), II, and IV) or be dedicated to single substrates (e.g., myosin light chain kinase, phosphorylase kinase, or CaMK III, now known as elongation factor-2 kinase).

Calmodulin-dependent protein kinases were among the first enzymes known to be regulated by calmodulin. In 1978, Cohen et al. identified CaM as the fourth subunit of the phosphorylase kinase [6], the groups of Hartshorne and Yagi described the calmodulin-dependent regulation of the myosin light chain kinase (MLCK [7, 8]), and Schulman and Greengard [9] reported on the Ca^{2+}-dependent protein phosphorylation and its activation by the calcium-dependent regulator, later identified as the calmodulin-dependent protein kinase II (CaMKII).

Calmodulin-dependent protein kinases are activated by phosphorylation like most protein kinases. This is achieved either by autophosphorylation or by another kinase. The latter was recently identified as a calmodulin-dependent calmodulin kinase kinase (CaMKK) [10, 11] suggesting that it would play a similar role to MAP-kinase kinases in the MAP kinase cascade.

In this chapter I will outline some of the structural features of the different calmodulin-dependent protein kinases, describe their different functional properties, and will evaluate their participation in different cellular regulatory pathways. Due to space limitations I cannot dwell on all aspects of this class of enzymes, but a more detailed analysis can be found in some of the earlier reviews on this subject (e.g., see [3, 4, 12–15]).

2
General Principles of Calmodulin-Dependent Protein Kinases

The different calmodulin-dependent protein kinases can be characterized by a segmental organization of the primary structure typical for most protein kinases. As schematically presented in Fig. 1, the enzymes are usually composed of a catalytic domain at the N-terminal site and of a regulatory domain at the C-terminal, and in some cases also of an association domain which either interacts with other subunits of an oligomeric complex or with intracellular targets. In the catalytic domain consensus sequences for serine/threonine protein kinases can be identified which are important for the activation of the enzyme, either by autophosphorylation and/or by phosphorylation of other kinase(s). The regulatory domain is divided into an autoinhibitory and a calmodulin binding domain which can be organized differently in the different kinases, i.e., either representing a contiguous sequence of the two subdomains which could even be overlapping each other, or residing on two separate peptides. The autoinhibitory domain may exert its action by binding in the active site, mimicking interactions with the catalytic center like the substrate, and is therefore often defined as a "pseudosubstrate" as discussed in detail by Kemp et al. [12]. Identification of autoinhibitory sites in calmodulin-dependent protein kinases has been obtained by site-directed mutagenesis studies rendering the kinase constitutively active without influencing the calmodulin binding property. A direct verification of this concept was obtained for another calmodulin-dependent enzyme, the calcium pump of plasma membranes, for which the receptor sites of the calmodulin binding domain were directly identified by cross-linking experiments [16, 17], and which were also responsible for its autoinhibition.

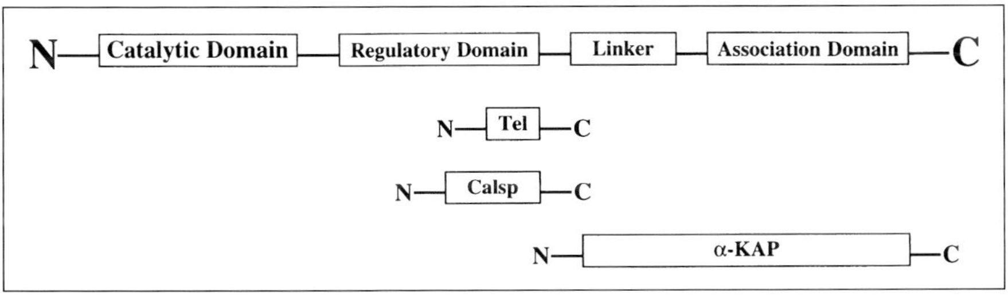

Fig. 1. Schematic representation of the domain composition of the different CaM-dependent protein kinases. The composition and length of the domains can vary between the different kinases (see text). The regulatory domain contains N-terminal the autoinhibitory and C-terminal the CaM-binding domain, which can be separate entities or overlap each other, dependent on the kinase. Linker and association domains can contain variable regions especially documented for CaMKII. Tel – telokin or KRP, kinase related protein, the internal gene product of MLCK represents the C-terminal part of the CaM-binding domain. Calsp = calspermin, gene product of CaMKIV contains the entire CaM-binding domain. α-KAP, the gene product of CaMKII, consists of the linker and association domain of CaMKII

3
Structural Properties of Calmodulin-Target Interactions

As discussed in detail by Yuan et al. in this volume, CaM undergoes a profound conformational change upon binding of Ca^{2+} to the different EF-hand loops of the protein. As a consequence deep hydrophobic clefts are built which are a prerequisite for the interaction with targets. In the absence of a target the N-terminal and the C-terminal halves of CaM behave as independent structural units in solution which are linked by a flexible tether. The latter is part of the seven turn central helix observed in the crystal structure of the protein [18].

CaM-binding sequences of different targets reveal similar properties (see Table 1). They are composed of short peptides with an average length of 20–30 amino acids which have the propensity to form a basic amphiphilic α-helix [19]. Several three-dimensional structures of complexes between calmodulin and target peptides representing binding domains of different CaM-dependent kinases, such as skeletal muscle [20] and smooth muscle MLCK [21], CaM-dependent protein kinase II [22], and CaM-dependent protein kinase kinase [23], have been solved by either NMR or X-ray crystallography. These structures share some common structural features:

1. Calmodulin, which usually has an extended, dumbbell-type shape, is collapsed to a globular structure wrapped around the target peptide.
2. The peptide is located in a hydrophobic channel passing through the center of the globular molecule.
3. Two hydrophobic residues of the binding peptide spaced by a defined number of residues are essential for the interaction with CaM.
4. The peptides interact with CaM in an antiparallel manner, i.e., the N-terminal half of CaM binds the C-terminal part of the peptide and vice versa.

An exception to the last rule was recently reported by Ikura and his coworkers [23] who solved the structure of the complex of CaM bound to the peptide of the binding domain of CaMKK by NMR. In this complex, the peptide comprises unusual structural properties, i.e., only the N-terminal half forms an α-helix – the C-terminal half is a hairpin loop folding back on itself. In addition, the peptide is oriented in a parallel fashion relative to the two halves of CaM.

Interestingly, recent experiments with small-angle X-ray scattering and neutron scattering provided evidence that when CaM is complexed to an intact enzyme, i.e., MLCK, CaM also undergoes a conformational collapse identical to that observed

Table 1. Aligned sequences of CaM-binding domains

skMLCK (M13)	KRR	[W]	KKNFIAVS	A	ANR	[F]		KKI
smMLCK (R20)	RRK	[W]	QKTGHAVR	A	IGR	[L]		SSM
CAMKI	KSK	[W]	KQKFNATA	[V]	VRH	[M]		RKL
CAMKII	RRK	[L]	KGAI LTTM	[L]	ATR	N		FSA
CAMKIV	RRK	[L]	KAAVKAVV	A	SSR	[L]		GSA
CAMKK	I PS	[W]	TTV I LVKS	M	LRK	R		S[F]G

Boxed hydrophobic residues represent the N-terminal and C-terminal anchors of the different CaM-binding domains.

with the peptide corresponding to its binding domain [24]. In addition, these experiments showed that a significant movement of its CaM-binding and autoinhibitory domains occurred due to the binding of CaM, away from the surface of the catalytic core to release inhibition of the enzymatic activity [24], postulated as a general mechanism for CaM-dependent enzymes. The first crystal structure of a CaM-dependent enzyme, CaM-dependent kinase I (CaMKI), obtained in the absence of CaM, i.e., in the autoinhibited state [25], supports this view in the sense that the C-terminal regulatory domain of CaMKI forms a helix-loop-helix segment that interferes with the two domains of the catalytic core, i.e., with the binding site of peptide substrates and with the ATP-binding site. An interesting feature of this structure is the striking observation that the N-terminal part of the CaM-binding domain is accessible for an initial interaction with CaM as discussed in detail below. This part comprises the loop region of the regulatory domain including the conserved tryptophan which provides the hydrophobic anchor for the C-terminal half of CaM. Furthermore, it was recently postulated that the interaction between the C-terminal half of CaM and the CaM-binding domain of a CaM-dependent enzyme is necessary – and sometimes already sufficient – to release the autoinhibited state of the enzyme [26]. This view was based on the three-dimensional NMR structure of the complex between CaM and the N-terminal part of the binding domain of the plasma membrane calcium pump, a view which is supported by recent SAXS studies on MLCK [27] indicating that initial binding of CaM to MLCK already occurs at substoichiometric Ca^{2+} concentrations, i.e., at about 2 mol of Ca^{2+}/mol of CaM.

4 Properties of Different CaM-Dependent Protein Kinases

As indicated before, calmodulin-dependent protein kinases can be divided into two categories:

1. Kinases dedicated to single substrates, here called monosubstrate kinases
2. Kinases dedicated to different substrates, here called multisubstrate kinases

4.1 Monosubstrate CaM-Kinases

4.1.1 *MLCK*

The most extensively studied calmodulin-dependent enzyme is the myosin light chain kinase (MLCK). This enzyme catalyzes the phosphorylation of a specific serine residue in the N-terminal portion of the myosin II regulatory light chains. Myosin II of higher organisms is a hexamer composed of a pair of heavy chains and pairs of regulatory (RLC) and essential (ELC) light chains. The amino terminus of the heavy chain forms a globular head containing the catalytic site of the ATPase, the light chain binding sites, and the actin interacting site. Basically, two types of MLCK exist which differ genetically, biochemically, and physiologically, and which are expressed in two different types of muscles: the MLCK of skeletal muscles (skMLCK), and of smooth muscles (smMLCK); the latter is also expressed in nonmuscle cells where it probably regulates the cytoskeleton which could be important for cell shape, cytokinesis, and cell motility [28–31].

The two different MLCK enzymes are encoded by single, but different, genes [32]. SkMLCK is restricted in its expression to striated muscle tissues. It is about 600 amino acids in length, corresponding to a molecular mass of 65 kDa [33, 34], and its CaM-binding domain was the first to be determined of the different CaM-dependent enzymes [35]. In contrast, the expression pattern and genomic organization of sm/nmMLCK is more complex [32]. This is due to alternative splicing which results in different isoforms with significantly different molecular weights, ranging between 150–200 kDa, due to different lengths of their amino termini [36]. The MLCK from non-muscle cells shows the longest amino terminal extension [37]. Two aspects are worth noting: (1) Non-muscle MLCK contains a modular motif of 12 amino acids near the N-terminus [38] which can be repeated up to 15 times and which has similarities to repeated motifs present in other proteins associated with chromatin structures. This is especially of interest since some MLCK isoform has been reported to be located in the nucleus [39]. (2) The carboxyl terminus of smMLCK is identical to a 20-kDa protein called either KRP for kinase related protein [32] or telokin [40]. This protein (see Fig. 1) can be expressed as an independent protein under the control of its own promoter which is located within the intron between the two exons encoding the CaM-binding domain of the smMLCK [41]. This is very much reminiscent of the organization of CaMKIV of which the C-terminal sequence can be expressed as a separate protein under the control of its own promoter located within an intron of the CaMKIV gene and which is expressed in a tissue specific manner [42] (see below). Telokin resembles an Ig-like motif which can be repeated up to four times in smMLCK and which is important for binding to myosin as was reported for titin [43]. It is further interesting to note that recently an embryonic form of non-muscle MLCK was reported with a molecular mass of 208 kDa which is downregulated after birth and which is distinct from other MLCK within their amino and carboxyl termini [44].

To date, the regulatory myosin light chain is the only established substrate of MLCK. In 1978 it had been shown that calmodulin mediates the Ca^{2+}-dependent activity of the enzyme [7, 8], at a time when calmodulin was primarily considered to be the Ca^{2+}-dependent activator of the cAMP phosphodiesterase. In smooth muscles the Ca^{2+}-dependent trigger of muscle contraction is directly mediated by the phosphorylation of the light chain by MLCK, thus removing the inhibition of the myosin ATPase [45]. In skeletal muscle, however, the acto-myosin ATPase is regulated by binding of Ca^{2+} to TroponinC, the regulatory calcium-binding protein in the Troponin complex. A possible function of MLCK in skeletal muscle is to modulate the rate of enhancement and extent of isometric contraction of the muscle as discussed by Sweeney et al. [46].

The substrate specificity of MLCK is very strict, and therefore the consensus sequence of the phosphorylation site of its substrate is much more complex compared to those of the multifunctional kinases such as CaMKII or IV. Kemp and Pearson [47], who studied the spatial requirements for location of the basic residues in peptide substrates of smMLCK, provided evidence that substrate recognition is basically dependent on the appropriate location of at least four basic residues over a more extended region N-terminal of the phosphorylated Ser residue. In the substrate recognition sequence of smMLC (11–22; K-K-R-P-Q-R-A-T-**S**-V-N-F) the residues at P – 8, P – 7, P – 6, and P – 3 have the strongest influence on the specificity which is also based on the spatial distance between the first three basic residues and Arg at P – 3 which should be separated by either two or three residues [47].

As indicated before, calmodulin-dependent enzymes are activated by disrupting an autoinhibitory segment which interacts with the catalytic core of the enzyme, thus deinhibiting the enzyme. This concept was first developed and verified for the cAMP-dependent protein kinase (PKA) which could be blocked by an endogenous inhibitor (PKI) competing with the substrate for the substrate binding site, thus serving as a pseudosubstrate [48]. MLCK is the CaM-dependent enzyme which has been studied very thoroughly to develop the pseudosubstrate hypothesis (for review see [49]). Kemp and his co-workers [50] reported an arrangement of basic residues within the CaM-binding domain of smMLCK which showed similarities to the consensus sequence of the substrate of the enzyme. Peptides corresponding to such a domain of either smMLCK or skMLCK inhibited constitutively active fragments of the MLCK, thus supporting the existence of an autoinhibitory domain in MLCK. This view gained further support from the finding that controlled proteolysis at the C-terminal end of MLCK generated a constitutively active fragment independent of calmodulin activation, i.e., the autoinhibitory domain was removed. On the other hand, subtle but significant differences in the activation pattern by CaM must exist between skMLCK and smMLCK, since the same CaM mutants activated the two homologous proteins with different efficiencies [51]. The mutants used for these experiments have been mutated systematically in the different Ca^{2+}-binding loops to impair Ca^{2+}-binding [51].

4.1.2
Phosphorylase Kinase

Phosphorylase kinase is the key regulatory enzyme in glycogen metabolism by converting the inactive phosphorylase **b** into the active phosphorylase **a** by phosphorylation which couples glycogenolysis with muscle contraction. Phosphorylase kinase is composed of four different subunits $\alpha\beta\gamma\delta$, each existing four times in the native complex, i.e., $(\alpha\beta\gamma\delta)_4$, summing up to a molecular mass of about 10^6 Da. The kinase activity is located within the γ-subunit, whereas the δ-subunit was identified as calmodulin [6] which is bound to the complex in a Ca^{2+}-independent manner. The tight interaction between the δ-subunit calmodulin and the rest of the enzyme may be due to two putative, closely spaced CaM-binding domains on the γ-subunit of phosphorylase kinase flanking an autoinhibitory domain [52] which are probably responsible for the high Ca^{2+} sensitivity of the enzyme. In addition, exogenous calmodulin can also interact with phosphorylase kinase, thereby activating the kinase even further.

Next to the activation of phosphorylase kinase by Ca^{2+}-mediated signal transduction, the enzyme is also activated by cAMP/PKA due to phosphorylation of two critical Ser residues in the β-subunit [53]. Phosphorylation of the α-subunit may also play a role in the activation of the holoenzyme [54].

The three-dimensional structure has been obtained for the catalytically active, but truncated form of the γ-subunit of phosphorylase kinase [55]. Even if the CaM-binding domains were missing there are some striking similarities to the structure of CaMKI in its autoinhibited form [25].

Evidence has accumulated that a number of isozymes exist for the different subunits of phosphorylase kinase which are encoded by different genes. The α-subunit is encoded by two different genes, mainly expressed in either muscle or liver, the latter isoform also in other nonmuscle tissues with several isotypes due to alternative splicing [56]. The β-subunit is encoded by one gene with several tissue-specific spliced isotypes [57]. The gene is 140 kb in length and is organized into 33 exons, 30 of which are expressed constitutively [57]. Exons 26 and 27 are mutually exclusive exons encoding either muscle or non-muscle type sequences. Three different genes can encode the γ-subunit which are tissue-specific expressed and are therefore called either γ_L for the liver type, γ_M for muscle, and γ_T for the testis type isoform [58], but it is possible that the liver and testis type isoforms result from the same gene. The muscle and liver/testis type isoforms have different tissue distributions, i.e., the muscle type is next to muscles mainly expressed in heart and brain, whereas the liver/testis type is mainly expressed in those tissues [58]. Calmodulin representing the δ-subunit is encoded by at least three separate genes in mammalian organisms [59], each encoding an identical protein, but it is not known which of these genes is linked specifically to the expression of the δ-subunit of the phosphorylase kinase.

4.1.3
CaMKIII or Elongation-Factor 2 Dependent Kinase

CaMKIII was originally identified as a calmodulin-dependent kinase which phosphorylates selectively elongation-factor 2 (eEF-2) [60–62]. After cloning and sequencing of the enzyme the primary structure revealed a protein of 724 amino acids in length and a molecular mass of 82 kDa [63] which was rather distantly related to other CaM-kinases and showed a low homology to known kinases.

CaMKIII phosphorylates two adjacent threonines in the N-terminal region of the elongation factor 2, Thr^{56} and Thr^{58} [62] in an ordered manner. This results in an inactivation of the factor [62, 64] due to abolishing its ability to interact with the ribosome [65]. The two threonine residues are located in a sequence not typical for a CaM-kinase consensus sequence, but the two residues are highly conserved during evolution, including *S. cerevisiae*, and yeast EF-2 is an excellent substrate for mammalian CaMKIII and vice versa [66]. The kinase is dependent on Ca^{2+}/calmodulin. The enzyme undergoes autophosphorylation and is thereby becoming independent of calmodulin, similar to CaMKII (see below) [66, 67].

Recently, Ryazanov et al. reported the cloning and sequencing of a number of CaMKIII homologues from different organisms including *Caenorhabditis elegans* [68]. By comparing these sequences with known members of other kinase families they could not identify sequence motifs typical for eucaryotic protein kinases with the exception of the ATP-binding site [68]. On the other hand, Ryazanov et al. identified a highly conserved region of about 200 amino acids with high homology to the catalytic domain of the recently described myosin heavy chain kinase A from *Dictyostelium discoideum* [69], and therefore suggested that CaMKIII define a novel subclass of protein kinases together with myosin heavy chain kinase A, with the consequence that CaMKIII should be named elongation factor-2 kinase.

4.2 Multisubstrate CaM-Kinases

Four different enzymes of calmodulin-dependent protein kinases fall into the class of multifunctional or multisubstrate kinases, CaMKI (V), CaMKII, CaMKIV, and CaMKK, the activating kinase of CaMKI and IV. This class can be subdivided into CaMKII on one side and CaMKI (V), CaMKIV, and CaMKK on the other for two reasons: (1) CaMKI (V), CaMKIV, and CaMKK act as monomeric proteins whereas CaMKII is a multimeric enzyme composed of up to 12 subunits; (2) CaMKI (V) and CaMKIV are activated by another calmodulin-dependent protein kinase, i.e., by CaMKK. Whether CaMKK itself is also activated by an upstream acting kinase is not clear at the moment, but there are indications that this indeed might be the case. In contrast, CaMKII is activated by autophosphorylation. Therefore the latter is discussed separately from CaMKI (V), CaMKIV, and CaMKK.

4.2.1 *CaMKII*

Calmodulin-dependent protein kinase II (CaMKII) represents a family of closely related protein kinases with a broad substrate specificity. This class of protein kinases is involved in cellular metabolism, neurotransmitter release, cytoskeletal structure, gene expression, cell cycle control, and – more recently – in long term potentiation and depression (LTP and LTD), functions which are thought to be involved in memory and learning.

CaMKII is highly abundant in the brain, and is especially enriched in postsynaptic densities [70]. Of the vertebrate CaMKII family four different isoforms exist, i.e., α, β, γ, δ, of molecular mass between 50 kDa and 60 kDa, which are derived from four different genes [3]. Further diversity is probably obtained by alternative splicing, although direct evidence exists only for the β-isoform [71, 72].

4.2.1.1 *Structural Organization of CaMKII*

The holoenzyme is an oligomeric complex of 6–12 subunits with a total molecular mass ranging from 300 kDa to 700 kDa which can be either a homomultimer or a heteromultimer. According to the amino acid sequences the different isoforms are closely related to each other and reveal the segmental organization typical for CaM-dependent protein kinases, consisting of catalytic, regulatory, variable, and association domains on the same polypetide chain (see Fig. 1). This is different from other kinases such as PKA or the cell cycle controlling kinases for which the catalytic and the regulatory domains are located on different proteins. Differences to other CaM-dependent kinases mainly exist outside the catalytic and the regulatory domains, e.g.,

CaMKII does not have such a pronounced N-terminal extension as some of the sm/nmMLCK, or it possesses an association domain downstream of the regulatory domain unlike CaMKI or IV, since the subunits of CaMKII have to assemble to build the oligomeric holoenzyme [3].

The variable domains differ in their length of sequence and their insertion location in the different isoforms. Of the 11 different types of variable domains (see Fig. 1), certainly the insert responsible for intracellular targeting of CaMKII to the nucleus [73, 74] is one of the most interesting domains since it enables CaMKII to participate in Ca^{2+}-dependent gene expression [75] (see also the discussion in Sect. 4.2.2.4). Nuclear targeting of CaMKII can be specifically blocked by phosphorylation of a serine residue adjacent to the nuclear localization sequence by either CaMKI or IV [76]. Similarly, a rudimentary gene product of CaMKIIα, α-KAP (see Fig. 1), consisting basically of the association domain of CaMKII, can be used to target CaMKII specifically to the sarcomeres of skeletal muscles where α-KAP serves as an anchoring protein for CaMKII [77, 78].

The overall structure of the catalytic domain of CaMKII is likely to be bilobal, as expected for all CaM-dependent protein kinases based on the crystal structure of CaMKI [25]. Catalysis is taking place within a cleft between the two lobes orienting the consensus sequence of the substrate in such a way that the basic residue(s) in proximity to the Ser/Thr residue which should be phosphorylated could interact with conserved polar residues located on the surface of the large lobe. The small lobe contains a lysine residue which is essential for ATP binding, i.e., if mutated to a methionine renders the kinase inactive [79].

4.2.1.2
Regulation of the Catalytic Activity

Since CaMKII has a broad range of substrates the consensus sequence for substrate determination is less strict than described for MLCK before. Basically, an Arg at the $P-3$ position – the phosphorylated Ser/Thr being P(0) – is the only essential requirement for a multifunctional kinase such as CaMKII [80]. In addition, hydrophobic residues are preferred at $P-5$ and $P+1$ [81].

Downstream of the catalytic domain the autoinhibitory, the CaM-binding, and the association domains are located, interspersed with variable domains. The latter give rise to a great variation in the composition of the different isoforms β, γ, δ due to alternative splicing, whereas spliced isoforms of α are currently unknown.

As discussed in detail above, CaM activates its targets by wrapping around the binding domain, thereby releasing the enzyme from its autoinhibitory state. In the case of CaMKII the three-dimensional structure of the complex between CaM and a peptide representing the CaM-binding domain of CaMKII has been determined by X-ray crystallography [22]. This complex also shows the general feature observed for similar CaM-peptide complexes [20, 21], i.e., CaM is collapsed to a globular structure engulfing the helical peptide which is fixed by numerous hydrophobic and electrostatic interactions. However, the hydrophobic residues critical for anchoring the CaMKII petide to the C-terminal and the N-terminal lobe of CaM are both leucine residues, Leu299 and Leu308, which are separated only by 8 amino acids instead of 12 as reported for the MLCK structures [20, 21]. Another important difference between the interactions of CaM with CaMKII and MLCK, respectively, is the different affinity of CaM for the two enzymes, i.e., the affinity for CaMKII is much lower (K_{act} 20–100 nmol/l) than for MLCK (K_{act} 1 nmol/l). But the affinity for CaMKII can be substantially enhanced (>1000-fold) by autophosphorylation of Thr286 resulting in a significant lowering of the CaM off-rate from the complex, a phenomenon termed "calmodulin trapping" [82] which might have some important implications for the function of the enzyme as discussed below.

The mechanism of activation of CaMKII is more complex compared to MLCK since the autophosphorylation site Thr286 is located within the autoinhibitory domain. Initial autophosphorylation of Thr286 is CaM-dependent resulting in a CaM-independent active enzyme which can modulate its activity by further, CaM-independent autophosphorylation [83, 84].

The autoinhibitory domain of CaMKII seems to be functionally different from MLCK since it fulfills two functions in the sense that it blocks not only the substrate site, i.e., being a "pseudosubstrate," but also the ATP binding site [85]. Phosphorylation of Thr286 within the autoinhibitory domain is not carried out by the catalytic site it is inhibiting, but by the catalytic site of a neighboring subunit of the holoenzyme [3]. Next to phosphorylation in the autoinhibitory site – the autonomy site – CaMKII can also be phosphorylated in the CaM-binding domain at Thr305 and Thr306 – the inhibitory sites, not only that CaM binding promotes inter-subunit phosphorylation of the autonomous site – it also protects the CaM-binding domain from being phosphorylated. One of the surprising consequences of autophosphorylation at Thr286 is the enormous increase in affinity for CaM, from about 50 nmol/l in the unphosphorylated state to 60 pmol/l after autophosphorylation [82].

The critical role of the phosphorylation of Thr286 for "trapping" CaM could be explained on the basis of the crystal structure of CaMKI [25]. The corresponding residue in CaMKI, Val290, is flanked by hydrophobic amino acids which are in contact with hydrophobic residues of the catalytic core. Due to the binding of CaM these contacts presumably are disrupted and Thr286 becomes accessible for phosphorylation. On the other hand, due to the dissociation of CaM from the kinase, the protein should relax back to the autoinhibited state by establishing again the hydrophobic contacts which is probably prevented by the negatively charged phosphate on Thr286 [25]. A recent detailed study by Yang and Schulman [86] using double mutants of critical residues in the autoinhibitory domain and/or their interacting partners in the catalytic domain seems to support this view.

4.2.1.3
CaMKII and Neuronal Plasticity

Cyclic cellular depolarization or hormonal stimuli can trigger a series of calcium pulses inside cells, either by Ca^{2+} influx through ligand-operated or voltage-sensitive channels or via IP_3-sensitive pathways. In neurons these Ca^{2+} oscillations can be used to modulate synaptic activity and strength with subsequently increased release or resynthesis of neurotransmitters or other cellular activities. Coordinator of at least some of these activities appears to be CaMKII since tyrosine hydroxylase and synapsin I, two well characterized substrates of CaMKII [87–90], play a critical role in neurotransmitter synthesis and release, respectively. The question now arises: Is CaMKII able to decode Ca^{2+} oscillations, and if so, how?

It has been known for some time that neurotransmission can be increased or decreased for a considerable length of time based on the history of synaptic activity which seemed to be dependent on CaMKII activity (for a review see [91]). Often hippocampus has been used to study such processes in which high frequency stimulation leads to long term potentiation (LTP) and low frequency to long term depression (LTD). These processes have been implied as important models to study spatial learning and memory storage [92]. Interestingly, Fukunaga et al. [93] reported that both processes are induced by an increase of intracellular Ca^{2+} in the postsynaptic neuron concomitant with a rise in CaMKII activity and phosphorylation of synapsin I [93, 94]. Further support for the involvement of CaMKII in LTP processes came from experiments with CaMKII inhibitory peptides which have been injected into appropriate cells and a block of LTP was observed by electrophysiological measurements, but not when control peptides have been used [95].

An important link has been made between LTP and some forms of learning, and its dependence on CaMKII by gene-targeted disruption of CaMKIIα by the group of Tonegawa [96, 97]. These studies provided important evidence not only that those mice which developed normally were impaired in spatial learning but also that induction of LTP was blocked in hippocampal slices from those mice.

An important – and elegantly designed – stop-flow experiment was recently reported by De Koninck and Schulman which provided direct evidence that CaMKII indeed can decode the frequency of Ca^{2+} pulses [98]. Due to the complex activation pattern of CaMKII by CaM, autophosphorylation and subsequent CaM trapping, the enzyme can become autonomous. As already pointed out, autophosphorylation is an inter-subunit reaction between proximate subunits in which CaM not only activates the "kinase" subunit, but also presents the "substrate" subunit for phosphorylation. Subsequently, the kinase is transformed into the "trapped" state, i.e., a cooperative, positive feedback loop resulting in a short molecular "memory" which could enable the enzyme to respond to digital and cyclic activation associated with Ca^{2+} oscillations. In simulation calculations, Hanson et al. [99] predicted that repetitive Ca^{2+} pulses lead to recruitment of CaM, autophosphorylation, and trapping of CaM, establishing a threshold frequency at which the activity of the enzyme is sustained [99]. These predictions were exactly confirmed by the experiments of De Koninck and Schulman [98] who could demonstrate that, independent of the Ca^{2+}-CaM pulse duration, the autonomous activation of CaMKII increased steeply as a function of frequency. On the other hand, once a threshold value was achieved it was possible to maintain the response level with signals of substantially lower frequency. These properties could be even more varied by the subunit composition as demonstrated by Brocke et al. [100] who studied heteromers of CaMKII composed of different ratios of α and β, since the affinity of the two isoforms for CaM differ substantially. On the other hand, Shen et al. [101] provided evidence that the β-isoform could be used to target α/β heteromeric holoenzymes to the actin cytoskeleton to localize CaMKII to synaptic sites of action. In summary, an important consequence of these results is the possibility that CaMKII not only can respond to a broad range of oscillation frequencies, either to rapid Ca^{2+} spikes as occurring during synaptic transmission or to slower, but broader waves as observed in nonexcitable cells, but also can be vectorially directed to places of action via the cytoskeleton.

4.2.2 *CaMKI and IV*

CaMKI and CaMKIV were both first isolated from rat brain, CaMKI as a synapsin I kinase [89, 102], CaMKIV from granular cells of the cerebellum [103]; therefore the latter enzyme was first named "CaMkinase Gr" [103]. In contrast to CaMKI, CaMKII, or calcineurin which are ubiquitously expressed, the tissue distribution of CaMKIV is restricted to nervous tissues [42], the thymus, particularly T-lymphocytes [104], the bone marrow, keratinocytes, testis, and ovary [166]. In all other tissues examined, CaMKIV remained undetectable [103, 105, 106]. Both CaMKI and IV belong to the class of multifunctional, multisubstrate CaM-dependent protein kinases, since both enzymes phosphorylate, next to synapsin I and II, a number of other substrates, mainly transcription factors (see below). CaMKI was first purified by Nairn and Greengard [107] as a 42-kDa monomeric protein of which several isoforms are known to date, including CaMKV which is highly homologous to CaMKI and is activated by the same CaMKK. Therefore CaMKI will be used as a name when generally referring to the different isoforms.

4.2.2.1 *The Crystal Structure of CaMKI*

In 1996 Goldberg et al. reported the crystal structure of CaMKI in the autoinhibited state, i.e., in the absence of CaM [25]. This is the first three-dimensional structure of a calmodulin-dependent enzyme determined at high resolution. Based on molecular modeling studies using the crystal structure of the cAMP protein kinase (PKA) as a template, it was proposed in 1992 [108] that the autoinhibitory site of MLCK could occupy the catalytic core of the enzyme. The crystal structure of CaMKI [25] provided evidence of significant contacts between the autoinhibitory sequence and the catalytic core of the kinase, thereby supporting the pseudosubstrate model for activation of calmodulin-dependent protein kinases. On the other hand, residues 300–316 of the regulatory segment comprising the main part of the CaM-binding domain do not directly enter the ATP-binding site, but interact with the outer part of the ATP-binding domain. This leads to a handle-like loop structure of residues 300–304 including tryptophan 303 on the surface of the protein, whereas residues 305–316 aligned between the ATP-binding β-strands have extended contacts to the outer surface of the ATP-binding domain. This arrangement, especially with the surface exposed, conserved tryptophan of the CaM-binding domain (e.g., see Fig. 7 of Yuan et al., this issue), makes it very likely that CaM could first bind to the exposed Trp 303 and subsequently release the β-strands of the ATP-binding domain from their interaction with the regulatory segment comprising the core of the CaM-binding domain [25]. The structure further suggests that binding of CaM and subsequent reorientation of the ATP-binding domain would open up the substrate binding domain which would permit Thr 177 to be accessible for the activating CaMKK. An interesting observation was recently reported by Chin and Means [109] who examined the basic requirements of activation of CaMKI by CaM based on the crystal structure of the enzyme just described. These authors observed that the CaM mutant M124Q resulted in about a 60-fold increase in the CaM-activation constant for CaMKI, suggesting that M124 of CaM is one of the major hydrophobic contacts with Trp 303 of the kinase. This view is also supported by the structures of the complex of CaM with the binding peptides of MLCK in which M124 is part of the hydrophobic cleft binding the conserved tryptophan [20, 21]. The importance of Trp 303 of CaMKI for the interaction with CaM and subsequent activation of the enzyme is further supported by the observation of Matsushita and Nairn [110] who reported that mutation of Trp303 to a serine residue reduced the affinity of CaMKI for CaM by a factor of 3000.

4.2.2.2
The Gene Structure of CaMKIV

CaMKIV is expressed in two spliced isoforms of M_r = 65,000 (α) and of M_r = 67,000 (β), of which the latter is found mainly in the cerebellum [111]. The two isoforms differ only by the presence of a 28-amino acid N-terminal extension on the β-isoform [112]. The primary structure has been determined for different mammalian species showing a high degree of similarity [42, 112–115]. The amino acid sequences of the regulatory and catalytic domains of CaMKIV are highly similar to corresponding regions of CaMKII, but the overall homology between the two enzymes is less than 50%. Recently the gene structure of CaMKIV has been reported [116]. The length of the gene is 42 kb of DNA, divided into 12 exons and 11 introns. A remarkable finding was that the last 169 amino acids of CaMKIV were identical to the entire coding sequence of another calmodulin-binding protein, calspermin, which is exclusively expressed in meiotic male germ cells of a number of mammalian species, but for which a function is unknown to date [42, 116]. The promoter region of calspermin could be identified in the tenth intron of the CaMKIV gene [116].

4.2.2.3
Regulation of Activity

The regulation of the activity of CaMKI and CaMKIV is a rather complex process and involves three major steps:

1. Binding of Ca^{2+}/CaM to the enzyme
2. Phosphorylation of either Thr177 (CaMKI) or Thr196 (CaMKIV) located in the activation loop of the kinases by CaMKK
3. Autophosphorylation of serine residue(s) at the N-terminus of the enzyme (only observed for CaMKIV)

If CaM is absent, CamKI and IV are locked in an autoinhibitory state like the other kinases discussed before. The boundary between the autoinhibitory segment and the calmodulin-binding domain is often not clearly defined, sometimes they are adjacent to each other, sometimes they are overlapping, but there are also cases in which the autoinhibitory sequence is part of the calmodulin-binding domain. With respect to CaMKIV, truncation experiments carried out in different laboratories seem to indicate that the autoinhibitory region minimally includes residues 305–317 [117, 118]. Adjacent to it, residues 319–335 can be identified as the calmodulin-binding domain which is highly homologous to other, well characterized calmodulin-binding sites (see Table 1), even if no structural data of complexes between CaM and the CaM-binding domains of CaMKI and IV are available to date. An interesting difference with respect to CaM-binding may exist between CaMKI and IV. In CaMKIV the two suggested anchoring residues – L322 and L335 – are spaced by 12 amino acids as in MLCK [20, 21]. On the other hand, in CaMKI the two anchoring residues could be W303 and V312 which would be separated by eight residues as in CaMKII [22] (see Table 1). This is in contrast to the suggestion made by Goldberg et al. [25] who proposed M316 as the second anchoring residue. Definite answers have to await appropriate structural studies. Another interesting difference between the two kinases concerns phosphorylation of a serine residue within the CaM-binding domain. Watanabe et al. [119] provided evidence that S332 is autophosphorylated by CaMKIV, which renders the kinase activity CaM independent. A homologous serine is lacking in CaMKI – it is replaced by a valine (see Table 1).

Difference in interaction between calmodulin and CaMKII on one hand and CaMKI and CaMKIV on the other could be reflected in the differences in their activation properties. As discussed before, CaMKII is a multisubunit, homo- or hetero-oligomer enzyme complex which is activated by autophosphorylation between two neighboring subunits [3, 120]. On the other hand, CaMKI and CaMKIV do not seem to multimerize, and both enzymes are activated by CaMKK; whether both are activated by the same kinase kinase or by different enzymes is not entirely clear. In CaMKIV, the critical residue phosphorylated by CaMKK is Thr196 which is located in the activation loop of the kinase [121, 122] and which is homologous to Thr177 in CaMKI. Accessibility of the two threonine residues is brought about by the binding of calmodulin, i.e., binding of calmodulin not only releases autoinhibition of I or IV, but also enables the enzymes to be phosphorylated at this critical residue. This phosphorylation results in a 10–20-fold activation in the case of CaMKIV [123, 124].

As indicated before, there is a third level of activation of CaMKIV concerning phosphorylation of serine residues at the N-terminus. In a detailed study, Chatila et al. [125] could demonstrate that Ser 11 and 12 of CaMKIV could be autophosphorylated in a slow process as indicated before, but CaMKK could considerably enhance this process by phosphorylating Thr196 [125].

Synthetic peptides helped to unravel the consensus sequences for CaMKI and IV which are important for substrate recognition and subsequent phosphorylation of either a serine or a threonine residue [118, 126]. These detailed studies revealed that CaMKI prefers hydrophobic residues at P – 5 and P + 4, and an Arg residue at P – 3, whereas for CaMKIV the only critical residue seems to be an Arg at position P – 3.

As indicated before, kinases – as well as phosphatases – play a critical role in many different regulatory pathways of the cell. Therefore it is very critical to keep a tightly controlled balance between the up- and down-regulation of these enzymes. In this context an important observation has recently been reported by Westphal et al. [127] who described the presence of a stable and stoichiometric complex between CaMKIV and the serine/threonine phosphatase 2A (PP2A) in activated T-lymphocytes. In this complex PP2A dephosphorylates CaMKIV and thus functions as a negative regulator of CaMKIV signaling. On the other hand, inhibition of PP2A leads to an enhanced CREB-mediated gene transcription by CaMKIV. Such a kind of tightly coupled signaling mechanism seems to be more common than expected since a similar complex has been described to exist between a MAP-kinase and the MAP-kinase phosphatase-3 [128].

4.2.2.4 *CaMKIV and Ca^{2+}-Dependent Gene Expression*

Early studies on Ca^{2+}-dependent gene expression indicated that CaMKI could play a critical role [75], but the participation of CaMKIV has been studied in much more detail, and therefore the latter will be discussed in this section.

One of the consequences of elevated calcium in the cell, especially in the nucleus, is the induction of gene expression [129]. Since transcription factors such as CREB [75, 130, 131], CREM τ [130, 132], ATF-1 [133], SRF [134], ETS-1 [135, 136] are among the best substrates for CaMKIV, and CaMKIV has been localized to the nucleus [130, 137, 138], the enzyme could have direct access to transcription factors to regulate their function in a Ca^{2+}-dependent manner. Thus it has been shown in different cell lines that CaMKIV is involved in the regulation of expression of immediate early genes (IEGs) either through CREB [130, 131, 139, 140] or through SRF [134].

Originally, CREB was identified by Gonzalez and Montminy [141] as a factor induced in a cAMP-dependent process which could be activated by PKA-dependent phosphorylation. Later studies provided evidence that this transcription factor could also be activated in a Ca^{2+}-dependent manner [75, 129] by either CaMKII or CaMKIV. Numerous studies demonstrated that all kinases activating CREB phosphorylated the transcription factor at the same serine residue: Ser133 [75, 130, 131, 140]. In contrast to CaMKIV, CaMKII phosphorylates next to Ser 133 also Ser 142, which has an inhibitory effect on gene expression as demonstrated by Sun et al. [140].

4.2.2.5 *Induction of CaMKIV During Development*

In a recent report, evidence was provided for the specific induction of CaMKIV by the thyroid hormone 3,3′,5-triiodo-L-thyronine (T_3) in a time- and concentration-dependent manner at a very early stage of brain differentiation [142]. In the immune system, in which CaMKIV is strongly expressed next to brain, it was also reported that the expression of CaMKIV could be hormonally induced [143]. In addition, it was interesting to observe that during T-cell differentiation the expression of CaMKIV was regulated. Thus, analysis of thymocyte subsets representing different developmental stages revealed that, during adult thymic T-cell development, the expression of CaMKIV was tightly controlled [143], indicating that CaMKIV was not expressed during the early stages of T-cell development, but started to be detectable first at the stage $CD25^{-}/44^{-}$. The highest concentration was observed in the $CD25^{+}/44^{+}$ cells [143], in agreement with observations reported by Hanissian et al. [104].

Maturation of T-cells in the thymus is a complex developmental process taking almost three weeks from a pluripotent stem cell to a fully developed mature $CD4^{+}$ or

$CD8^+$ SP cell during which time lineage commitment, gene rearrangement, expression of the T-cell receptor, and selection processes take place. Anderson et al. [144] recently demonstrated that transgenic mice expressing a catalytically inactive form of CaMKIV not only were hampered in T-cell activation, but also exhibited a significant defect in thymic cellularity, indicating that CaMKIV may play a critical role in T-cell maturation. Therefore, the authors discussed the possibility that defective CaMKIV expression could prevent the maturation of DN into DP cells. In this respect it is interesting to note that in a recent communication $CREB^{-}/^{-}$ mice were reported to exhibit a defect in T-cell development [145] which also showed a characteristic reduction in thymic cellularity. Thymocytes of these CREB deficient cells accumulate at the immature double negative stage ($CD25^{-}44^{-}$), unable to progress into the double positive stage ($CD25^{+}44^{+}$), i.e., the transition during which CaMKIV reaches its highest expression during T-cell development [143]. Thus it appears that the transcription factor CREB and its activating kinase CaMKIV are essential elements for T-cell development.

4.2.3
CaM-Dependent Protein Kinase Kinase (CaMKK)

As indicated above, CaMKI and IV are activated by calmodulin-dependent protein kinase(s), the CaMKK, as first determined by DeRemer et al. [10, 11]. Two isoforms of CaMKK are known to date, α and β, which are encoded by different genes [15]. Activation of CaMKI and CaMKIV by CaMKK is accomplished through a Ca^{2+}-CaM dependent phosphorylation of threonine residues in the activation loops of CaMKI and IV, Thr177 of CaMKI, and Thr196 of CaMKIV, respectively. αCaMKK has been cloned and characterized by Tokumitsu et al. [122] whereas βCaMKK was described by Kitani et al. [146] and by Anderson et al. [147], respectively. Both isoforms are organized similar to other CaM-dependent kinases (see Fig. 1) with catalytic, autoinhibitory, and CaM-binding domains. In addition, CaMKK contains an unusual Arg-Pro rich insert within the N-terminal part of its catalytic domain (see Fig. 1) which might be important for the recognition of CaMKI and IV as substrates, since deletion of these inserts abolishes its activation potential towards CaMKI and IV [148].

CaMKK is mainly present in brain, thymus, spleen, and testis, but can also be identified in other tissues at lower levels [15]. The enzyme can be found both in the cytoplasm and in the nucleus, the latter being important for the activation of CaMKIV which is mainly responsible for the Ca^{2+}-dependent gene expression [75]. In this context it is interesting to note that the distribution of βCaMKK in the brain is much more similar to CaMKIV, and that of αCaMKK follows the distribution of CaMKI [147], suggesting that βCaMKK could be the actual activator of CaMKIV, whereas αCaMKK would be that of CaMKI, respectively [147]. Nevertheless, both isoforms of CaMKK can activate CaMKI and IV with the same efficiency, but whether there exist more specific isoforms and whether there are additional kinases further upstream as in the MAP-kinase cascade is not known to date.

Interference with other second messenger pathways also exists since CaMKK can be phosphorylated in its CaM-binding domain by PKA like other CaM-dependent enzymes, thereby inhibiting its activity [149]. Another possible crosstalk is suggested by recent findings by Yano et al. [150] who reported that CaMKK could directly phosphorylate and activate protein kinase B which can inhibit apoptosis through phosphorylation of BAD, a pro-apototic member of the BCL-2 family, a possible new avenue which links the CaM-kinase cascade to apoptotic events.

4.3
Undefined CaM-Kinases

Recently, a number of CaM-dependent protein kinases have been identified of which the substrate specificity is not known unambiguously.

4.3.1
DAP-Kinase

A new member of the calmodulin-dependent protein kinases was detected recently. By searching genes which are specifically linked to programmed cell death (PCD) several laboratories introduced a functional selection strategy of gene cloning (see [151, 152]). Using this approach, several new "death associated proteins" (DAP) could be identified upon treatment of cells with interferon-γ [153], one of which, DAP-2, turned out to be a calmodulin-dependent protein kinase. DAP-2 is structurally unique, i.e., it is a 160-kDa serine/threonine protein kinase containing a number of other domains such as ankyrin repeats, P-loops, CaM-binding domain, and cytoskeleton-binding- and death-domains [153]. It was demonstrated that the enzyme could phosphorylate myosin regulatory light chains in a calmodulin-dependent manner, in addition to the activation by autophosphorylation. Deletion of the CaM-binding domain renders a constitutively active kinase, typical for CaM-dependent enzymes. Cellular localization experiments associated the enzyme with the cytoskeleton. The latter observation could be of special interest since the cytoskeleton organization preceded the nuclear condensation due to changes in actin microfilament, and segmentation occurred in response to interferon-γ-stimulation or DAP-kinase overexpression [153]. By ectopic expression in HeLa cells DAP-kinase induced cell death, whereas expression of a catalytically inactive mutant in HeLa cells protected these cells from interferon-γ-induced cell death. On the other hand, it was equally important that the catalytically active enzyme was localized correctly on the cytoskeleton. If a truncated but catalytically active form of the DAP-kinase was overexpressed in HeLa cells, which was mislocalized to the nucleus, this enzyme failed to disrupt the actin microfilament, and therefore the cells were not killed.

DAP-Kinase is widely expressed in many cells and seems to play a central role in apoptotic processes since a number of stimuli leading to PCD converge in the activation of DAP-kinase. A further, very interesting aspect of the role of this enzyme in cellular processes is its possible function as a tumor suppressor gene, indicated by the fact that the DAP-kinase could not be detected in a number of lymphomas, leukemia cell lines, and cell lines derived from a number of different carcinomas, albeit its wide expression in all human and murine tissues tested to date [154]. In addition, there could be a correlation between the metastatic property of carcinoma cells and DAP-kinase activity, since it was found that high-metastatic lung carcinoma clones lacked DAP-kinase expression whereas their low-metastatic counterparts expressed normal levels of this enzyme [155]. It was then striking to observe that, by restoring normal levels of DAP-kinase in highly metastatic Lewis carcinoma cells, these cells lost their ability to induce lung metastases after intravenous injection into mice [155].

4.3.2
CaM-Kinase of Plants

Calcium signaling can also be observed in higher plants. Early on, calmodulin was identified as an activator of the NAD-kinase from plants [156, 157]. In the course of characterizing CaM-dependent kinases in plants a novel class of calcium-dependent protein kinases was identified in which a calcium-binding domain homologous to calmodulin was fused to a catalytic domain of a serine/threonine kinase [158]. The first member of this novel class was purified from soybean cell cultures [159] soon after the first primary structure was reported [160]. The monomeric enzyme with an apparent molecular weight of 57,000 requires direct binding of Ca^{2+} for activation through a regulatory domain directly connected to the catalytic domain. The latter contains characteristic features of a catalytic domain, typical for conventional serine/threonine kinases including a region with characteristics of an autoinhibitory domain. This region is highly homologous to the corresponding domain of CaMKII. C-terminal to the catalytic domain is the regulatory domain which is highly homologous to calmodulin including four canonical helix-loop-helix domains of the EF-hand family which binds calcium with high affinity. Since these enzymes can be inhibited by positively charged amphipathic drugs and peptides in a calcium-dependent manner [161, 162], similar to calmodulin-dependent enzymes of higher organisms, it can be proposed that the calmodulin-like regulatory domain could interact with the autoinhibitory domain in a calcium-dependent manner to release autoinhibition. Normally these enzymes are soluble and are localized in the cytosol, but there are also reports of membrane-bound forms [163, 164] which can only be removed by detergent solubilization [164]. Endogenous substrates have not been identified unambiguously, but several lines of evidence indicate that this type of kinase also belongs to the multifunctional type [158]. One of the best characterized possible substrates is nodulin 26 [163], an integral membrane channel protein [165], but further detailed studies are necessary to unravel the role of this novel class of kinases in the plant kingdom.

5
Conclusions

In this review, structural properties and functional characteristics of the different calmodulin-dependent protein kinases have been summarized. It was outlined that the different kinases can mediate cellular responses to a variety of Ca^{2+} messages, ranging from sustained levels of calcium to calcium oscillations. Transfer of messages is regulated by either phosphorylating dedicated protein substrates by kinases such as MLCK or by coordinating the phosphorylation of a variety of different protein substrates by multifunctional CaMkinases. The activity of the different kinases can be controlled on different levels and by different means such as autophosphorylation or activation by kinase cascades. Mutual control of the different second messenger systems makes it possible for the cell to respond to a host of integrated messages in a differentiated manner. Future experiments will hopefully unravel details of the mechanisms of this tightly controlled network.

References

1. Ikura M (1996) Trends Biochem Sci 21:14
2. Cheung WY (1980) Science 207:19
3. Hanson PI, Schulman H (1992) Ann Rev Biochem 61:559
4. Schulman H (1993) Curr Opin Cell Biol 5:247
5. Klee CB, Ren H, Wang X (1998) J Biol Chem 273:13,367
6. Cohen P, Burchell A, Foulkes JG, Cohen PTW, Vanaman TC, Nairn AC (1978) FEBS Lett 92:287
7. Dabrowska R, Aromatorio D, Sherry JMF, Hartshorne DJ (1978) Biochemistry 17:253
8. Yagi K, Yazawa M, Kakiuchi S, Ohshima M, Uenishi K (1978) J Biol Chem 253:1338
9. Schulman H, Greengard P (1978) Proc Natl Acad Sci USA 75:5432
10. DeRemer MF, Saeli RJ, Edelman AM (1992) J Biol Chem 267:13,460
11. DeRemer MF, Saeli RJ, Brautigan DL, Edelman AM (1992) J Biol Chem 267:13,466
12. Kemp BE, Faux MC, Means AR, House C, Tiganis T, Hu S-H, Mitchelhill KI (1994) Structural aspects: pseudosubstrate and substrate interactions. In: Woodgett JR (ed) Protein kinases. Oxford University Press, New York, p 30
13. Kemp BE, Barden JA, Kobe B, House C, Parker MW (1996) Adv Pharmacol 36:221
14. Lukas TJ, Mirzoeva S, Watterson DM (1998) Calmodulin-regulated protein kinases. In: VanEldik L, Watterson DM (eds) Calmodulin and signal transduction. Academic Press, New York, p 65
15. Soderling TR (1999) Trends Biochem Sci 24:232
16. Falchetto R, Vorherr T, Brunner J, Carafoli E (1991) J Biol Chem 266:2930
17. Falchetto R, Vorherr T, Carafoli E (1992) Protein Sci 1:1613
18. Babu YS, Sack JS, Greenhough TJ, Bugg CE, Means AR, Cook WJ (1985) Nature 315:37
19. O'Neil KT, DeGrado WF (1990) Trends Biochem Sci 15:59
20. Ikura M, Clore GM, Gronenborn AM, Zhu G, Klee CB, Bax A (1992) 256:632
21. Meador WE, Means AR, Quiocho FA (1992) Science 257:1251
22. Meador WE, Means AR, Quiocho FA (1993) Science 263:1718
23. Osawa M, Tokumitsu H, Swindells MB, Kurihara H, Orita M, Shibanuma T, Furuya T, Ikura M (1999) Nature Struct Biol 6:819
24. Krueger JK, Olah GA, Rokop SE, Zhi G, Stull JT, Trewhella J (1997) Biochemistry 36:6017
25. Goldberg J, Nairn AC, Kuriyan J (1996) Cell 84:875
26. Elshorst B, Hennig M, Foersterling H, Diener A, Maurer M, Schulte P, Schwalbe H, Griesinger C, Krebs J, Schmid H, Vorherr T, Carafoli E (1999) Biochemistry 38:12,320
27. Krueger JK, Bishop NA, Blumenthal DK, Zhi G, Beckingham K, Stull JT, Trewhella J (1998) 37:17,810
28. Shoemaker MO, Lau W, Shattuck RL, Kwiatkowski AP, Matrisian PE, Guerra-Santos L, Wilson E, Lukas TJ, VanEldik LJ, Watterson DM (1990) J Cell Biol 111:1107
29. Karess RE, Chang XJ, Edwards KA, Kulkarni S, Aguilera I, Kiehart DP (1991) Cell 65:1177
30. Fishkind DJ, Cao LG, Wang YL (1991) J Cell Biol 114:967
31. Wilson AK, Gorgas G, Claypool WD, de Lanerolle P (1991) J Cell Biol 114:277
32. Collinge M, Matrisian PE, Zimmer WE, Shattuck RL, Lukas TJ, VanEldik LJ, Watterson DM (1992) Mol Cell Biol 12:2359
33. Takio K, Blumenthal DK, Edelman AM, Walsh KA, Krebs EG, Titani K (1985) Biochemistry 24:6028
34. Takio K, Blumenthal DK, Walsh KA, Titani K, Krebs EG (1986) Biochemistry 25:8049
35. Blumenthal DK, Takio K, Edelman AM, Charbonneau H, Titani K, Walsh KA, Krebs EG (1985) Proc Natl Acad Sci 82:3187

36. Watterson DM, Collinge M, Lukas TJ, Van Eldik LJ, Birukov KG, Stepanova OV, Shirinsky VP (1995) FEBS Lett 373:217
37. Potier MC, Chelot E, Pekarsky Y, Gardiner K, Rossier J, Turnell WG (1995) Genomics 29:562
38. Gallagher PJ, Herring BP, Griffin SA, Stull JT (1991) J Biol Chem 266:23,936
39. Simmen RCM, Dunbar BS, Guerriero V, Chafouleas JG, Clark JH, Means AR (1994) J Cell Biol 99:588
40. Ito M, Dabrowska R, Guerriero V Jr, Hartshorne DJ (1989) J Biol Chem 264:13,971
41. Gallagher PJ, Herring BP (1991) J Biol Chem 266:23,945
42. Means AR, Cruzalegui F, LeMagueresse B, Needleman DS, Slaughter GR, Ono T (1991) Mol Cell Biol 11:3960
43. Labeit S, Gautel M, Lakey A, Trinick J (1992) EMBO J 11:1711
44. Gallagher PJ, Garcia JGN, Herring BP (1995) J Biol Chem 270:29,090
45. Walsh MP (1994) Mol Cell Biochem 135:21
46. Sweeney HL, Bowman BF, Stull JT (1993) Am J Physiol 264:C1085
47. Kemp BE, Pearson RB (1985) J Biol Chem 260:3355
48. Demaille JG, Peters KA, Fischer EH (1977) Biochemistry 16:3080
49. Kemp BE, Faux MC, Means AR, House C, Tiganis T, Hu S-H, Mitchelhill KI (1994) Structural aspects: pseudosubstrate and substrate interactions. In: Woodgett JR (ed) Protein kinases. Oxford University Press, New York, p 30
50. Kemp BE, Pearson RB, Guerriero V, Bagchi IC, Means AR (1987) J Biol Chem 262:2542
51. Gao ZH, Krebs J, VanBerkum MFA, Tang W-J, Maune JF, Means AR, Stull JT, Beckingham K (1993) J Biol Chem 268:20,096
52. Dasgupta M, Honeycutt T, Blumenthal DK (1989) J Biol Chem 264:17,156
53. Yeaman SJ, Cohen P (1975) Eur J Biochem 51:93
54. Pickett-Gies CA, Walsh DA (1985) J Biol Chem 260:2046
55. Owen DJ, Noble ME, Garman EF, Papageorgiou AC, Johnson LN (1995) Structure 3:467
56. Davidson JJ, Ozcelik T, Hamacher C, Willems PJ, Francke U, Kilimann MW (1992) Proc Natl Acad Sci USA 89:2096
57. Wullrich-Schmoll A, Kilimann MW (1996) Eur J Biochem 238:374
58. Calalb MB, Fox DT, Hanks SK (1992) J Biol Chem 267:1455
59. Fischer R, Flura M, Strehler-Page M-A, Matthews S, Krebs J, Penniston JT, Carafoli E, Strehler EE (1988) J Biol Chem 263:17,055
60. Palfrey HC (1983) FEBS Lett 157:183
61. Nairn AC, Bhagat B, Palfrey HC (1985) Proc Natl Acad Sci USA 82:7939
62. Nairn AC, Palfrey HC (1987) J Biol Chem 262:17,299
63. Redpath NT, Price NT, Proud CG (1996) J Biol Chem 271:17,547
64. Redpath NT, Price NT, Severinov KV, Proud CG (1993) Eur J Biochem 213:689
65. Carlberg U, Nilsson A, Nygard O (1990) Eur J Biochem 191:639
66. Mitsui K, Brady M, Palfrey HC, Nairn AC (1993) J Biol Chem 268:13,422
67. Redpath NT, Proud CG (1993) Eur J Biochem 212:511
68. Ryazanov AG, Ward MD, Mendola CE, Pavur KS, Dorovkov MV, Wiedmann M, Erdjument-Bromage H, Tempst P, Parmer TG, Prostko CR, Germino FJ, Hait WH (1997) Proc Natl Acad Sci USA 94:4884
69. Futey LM, Medley QG, Cote GP, Egelhoff TT (1995) J Biol Chem 270:523
70. Erondu NE, Kennedy MB (1985) J Neurosci 5:3270
71. Bulleit RF, Bennet MK, Molloy SS, Hurley JB, Kennedy MB (1988) Neuron 1:63
72. Karls U, Muller U, Gilbert DJ, Copeland NG, Jenkins NA, Harbers K (1992) Mol Cell Biol 12:3644
73. Srinivasan M, Edman C, Schulman H (1994) J Cell Biol 126:839
74. Brocke L, Srinivasan M, Schulman H (1995) J Neurosci 15:6797
75. Sheng M, Thompson MA, Greenberg ME (1991) Science 252:1427
76. Heist EK, Srinivasan M, Schulman H (1998) J Biol Chem 273:19,763
77. Bayer KU, Lohler J, Harbers K (1996) Mol Cell Biol 16:29
78. Bayer KU, Harbers K, Schulman H (1998) EMBO J 17:5598
79. Hanks SK, Quinn AM (1991) Methods Enzymol 200:38
80. Kennelly PJ, Krebs EG (1991) J Biol Chem 266:15,555
81. Stokoe D, Caudwell B, Cohen PTW, Cohen P (1993) Biochem J 296:843
82. Meyer T, Hanson PI, Stryer L, Schulman H (1992) Science 256:1199

83. Lou LL, Schulman H (1989) J Neurosci 9:2020
84. Hanson PI, Schulman H (1992) J Biol Chem 267:17,216
85. Smith MK, Colbran RJ, Brickley DA, Soderling TR (1992) J Biol Chem 267:1761
86. Yang E, Schulman H (1999) J Biol Chem 274:26,199
87. Griffith LC, Schulman H (1988) J Biol Chem 263:9542
88. Waymire JC, Johnston JP, Hummer-Lickteig K, Lloyd A, Vigny A, Craviso GL (1988) J Biol Chem 263:12,439
89. Huttner WB, DeGennaro LJ, Greengard P (1981) J Biol Chem 256:1482
90. Czernik AJ, Pang DT, Greengard P (1987) Proc Natl Acad Sci USA 84:7518
91. Soderling TR (1995) Adv Second Mess Phosphoprotein Res 30:175
92. Bliss TVP, Collingridge GL (1993) Nature 361:31
93. Fukunaga K, Stoppini L, Miyamoto E, Muller D (1993) J Biol Chem 268:7863
94. Fukunaga K, Muller D, Miyamoto E (1995) J Biol Chem 270:6119
95. Malinow R, Schulman H, Tsien RW (1989) Science 245:862
96. Silva AJ, Stevens CF, Tonegawa S, Wang Y (1992) Science 257:201
97. Silva AJ, Paylor R, Wehner JM, Tonegawa S (1992) Science 257:206
98. De Koninck P, Schulman H (1998) Science 279:227
99. Hanson PI, Meyer T, Stryer L, Schulman H (1994) Neuron 12:943
100. Brocke L, Chiang LW, Wagner PD, Schulman H (1999) J Biol Chem 274:22,713
101. Shen K, Teruel MN, Subramanian K, Meyer T (1999) Neuron 21:593
102. Kennedy MB, Greengard P (1981) Proc Natl Acad Sci USA 78:1293
103. Ohmstede C-A, Jensen KF, Sahyoun NE (1989) J Biol Chem 264:5866
104. Hanissian H, Frangakis M, Bland MM, Jawahar S, Chatila TA (1993) J Biol Chem 268:20,055
105. Frangakis MV, Ohmstede C-A, Sahyoun N (1991) J Biol Chem 266:11,309
106. Sakagami H, Watanabe M, Kondo H (1992) Mol Brain Res 16:20
107. Nairn AC, Greengard P (1987) J Biol Chem 262:7273
108. Knighton DR, Pearson RB, Sowadski JM, Means AR, Eyck LFT, Taylor SS, Kemp BE (1992) Science 258:130
109. Chin D, Means AR (1996) J Biol Chem 271:30,465
110. Matsushita M, Nairn AC (1998) J Biol Chem 273:21,473
111. Frangakis MV, Chatila T, Wood ER, Sahyoun NE (1991) J Biol Chem 266:17,592
112. Sakagami H, Kondo H (1993) Mol Brain Res 19:215
113. Ohmstede C-A, Bland MM, Merrill BM, Sahyoun NE (1991) Proc Natl Acad Sci USA 88:5784
114. Jones DA, Glod J, Wilson-Shaw D, Hahn WE, Sikela JM (1991) FEBS Lett 289:105
115. Bland MM, Monroe RS, Ohmstede C-A (1994) Gene 142:191
116. Sun Z, Means RL, LaMagueresse B, Means AR (1995) J Biol Chem 270:29,507
117. Tokumitsu H, Brickey DA, Glod J, Hidaka H, Sikela J, Soderling TR (1994) J Biol Chem 269:28,640
118. Cruzalegui FH, Means AR (1993) J Biol Chem 268:26,171
119. Watanabe S, Okuno S, Kitani T, Fujisawa H (1996) J Biol Chem 271:6903
120. Braun AP, Schulman H (1995) Annu Rev Physiol 57:417
121. Selbert MA, Anderson KA, Huang QH, Goldstein EG, Means AR, Edelman AM (1995) J Biol Chem 270:17,616
122. Tokumitsu H, Enslen H, Soderling TR (1995) J Biol Chem 270:19,320
123. Tokumitsu H, Soderling TR (1996) J Biol Chem 271:5617
124. Tokumitsu H, Wayman GA, Muramatsu M, Soderling TR (1997) Biochemistry 36:12,823
125. Chatila T, Anderson KA, Ho N, Means AR (1996) J Biol Chem 271:21,542
126. Lee JC, Kwon YG, Lawrence DS, Edelman AM (1994) Proc Natl Acad Sci 91:6413
127. Westphal RS, Anderson KA, Means AR, Wadzinski BE (1998) Science 280:1258
128. Camps M, Nichols A, Gillieron C, Antonsson B, Muda M, Chabert C, Boschert U, Arkinstall S (1998) Science 280:1262
129. Sheng M, McFadden G, Greenberg ME (1990) Neuron 4:571
130. Matthews RP, Guthrie CR, Wailes LM, Zhao X, Means AR, McKnight GS (1994) Mol Cell Biol 14:6107
131. Enslén H, Sun P, Brickey D, Soderling SH, Klamo E, Soderling TR (1994) J Biol Chem 269:15,520
132. Sun Z, Sassone-Corsi P, Means AR (1995) Mol Cell Biol 15:561

133. Sun P, Lou L, Maurer RA (1996) J Biol Chem 271:3066
134. Miranti CK, Ginty DD, Huang G, Chatila T, Greenberg ME (1995) Mol Cell Biol 15:3672
135. Pognonec P, Boulokos KE, Gesquiere J, Stehelin D, Ghysdael J (1988) EMBO J 7:977
136. Fisher CL, Ghysdael J, Cambier JC (1991) J Immunol 146:1743
137. Jensen KF, Ohmstede C-A, Fisher RS, Sahyoun N (1991) Proc Natl Acad Sci USA 88:2850
138. Bito H, Deisseroth K (1996) Cell 87:1203
139. Enslén H, Soderling TR (1994) J Biol Chem 269:20,872
140. Sun P, Enslén H, Myung PS, Maurer RA (1994) Gen Dev 8:2527
141. Gonzalez GA, Montminy MR (1989) Cell 59:675
142. Krebs J, Means RL, Honegger P (1996) J Biol Chem 271:11,055
143. Krebs J, Wilson A, Kisielow P (1997) Biochem Biophys Res Commun 241:383
144. Anderson KA, Ribar TJ, Illario M, Means AR (1997) Mol Endocrinol 11:725
145. Rudolph D, Tafuri A, Gass P, Haemmerling GJ, Arnold B, Schuetz G (1998) Proc Natl Acad Sci USA 95:4481
146. Kitani T, Okuno S, Fujisawa H (1997) J Biochem (Tokyo) 122:243
147. Anderson KA, Means RL, Huang QH, Kemp BE, Goldstein EG, Selbert MA, Edelman AM, Fremeau RT, Means AR (1998) J Biol Chem 273:31,880
148. Tokumitsu H, Takahashi N, Eto K, Yano S, Soderling TR, Muramatsu M (1999) J Biol Chem 274:15,803
149. Wayman GA, Tokumitsu H, Soderling TR (1997) J Biol Chem 272:16,073
150. Yano S, Tokumitsu H, Soderling TR (1998) Nature 396:584
151. Deiss LP, Kimchi A (1991) Science 252:117
152. Kimchi A (1998) Biochim Biophys Acta 1377:F13
153. Cohen O, Feinstein E, Kimchi A (1997) EMBO J 16:998
154. Kissil J, Feinstein E, Cohen O, Jones PA, Tsai YC, Knowles MA, Eydmann ME, Kimchi A (1997) Oncogene 15:403
155. Inbal B, Cohen O, Polak-Charkon S, Kopolovic J, Vadai E, Eisenbach L, Kimchi A (1997) Nature 390:180
156. Muto S, Miyachi S (1977) Plant Physiol 59:55
157. Anderson JM, Cormier MJ (1978) Biochem Biophys Res Commun 84:595
158. Roberts DM (1993) Curr Op Cell Biol 5:242
159. Putnam-Evans CL, Harmon AC, Cormier MJ (1990) Biochemistry 29:2488
160. Harper JF, Sussman MR, Schaller EG, Putnam-Evans C, Charbonneau H, Harmon AC (1991) Science 252:951
161. Polya GM, Nott R, Klucis E, Minichello J, Chandra S (1990) Biochim Biophys Acta 1037:259
162. Zhang XQ, Chollet R (1997) FEBS Lett 410:126
163. Weaver CD, Crombie B, Stacey G, Roberts DM (1991) Plant Physiol 95:222
164. Schaller EG, Harmon AC, Sussman MR (1992) Biochemistry 31:1721
165. Pao GM, Wu L-F, Johnson KD, Hofte H, Chrispeels MJ, Sweet G, Sandal NN, Saier MH Jr (1991) Mol Microbiol 5:33
166. Means AR (1999) Proc 11th Int Symp Calcium Binding Prot, Kisarazu, Japan

Structure of Calcineurin and Its Complex with Immunophilins

Claude B. Klee

Laboratory of Biochemistry, National Cancer Institute, National Institutes of Health,
Bethesda, MD 20892 USA
E-mail: ckl@helix.nih.gov

Over the past ten years the elucidation of the mechanism of action of the immunosuppressive drugs, cyclosporin A and FK506, has helped us to understand a major problem in biology at the molecular level: how are external signals acting on plasma membrane receptors transmitted to the nucleus. The specific inhibition by immunosuppressive drugs, complexed with their binding proteins, of the Ca^{2+}/calmodulin-dependent protein phosphatase, calcineurin, prevents the dephosphorylation and translocation to the nucleus of a family of transcription factors regulating the expression of many genes.

Keywords. Calcium signaling, Calcineurin, Protein phosphatase, FKBP12, FK506, CsA, NFAT

1 Introduction

The discovery, more than twenty years ago, of the immunosuppressive activity of the cyclic peptide, cyclosporin A (CsA), isolated from the fungus *Tolypocladium inflatum*, revolutionized human organ transplants and the treatment of auto immune diseases [1]. The search for a CsA receptor led Handschumacher et al. [2] to isolate a 17,000 MW protein, cyclophilin A (CyPA) later identified as a peptidylprolyl *cis-trans* isomerase (PPIase) [3, 4]. As CsA was a potent inhibitor of the PPIase activity of CyPA it was tempting to conclude that the immunosuppressive activity of CsA was the result of the PPIase inhibition. The subsequent isolation of two fungal macrolides, FK506 (tacrolimus) produced by *Streptomyces tsukubaensis*, and rapamycin by *Streptomyces hygroscopicus*, with even more potent immunosuppressive activities [5, 6], and the demonstration that their cellular receptor, FKBPs for FK-Binding Proteins, belongs to a different class of peptidylprolyl *cis-trans* isomerases inhibited by FK506 and rapamycin, added support to this hypothesis [7, 8]. Because of their common involvement in immunosuppression CyPs and FKBPs are usually referred to as immunophilins. During the past ten years the combined use of synthetic chemistry, molecular biology, biochemistry, cell biology, structural biology and yeast genetics, has led to the demonstration that the immunosuppressive properties and cytotoxic effects of these fascinating drugs are the result of the formation of a toxic complex with their binding proteins rather than of the inhibition of the PPIase activity of their binding proteins [9 and ref. therein].

2 Inhibition of the PPIase Activity and Immunosuppressive Effects of CsA and FK506

The two structurally related, but functionally different, immunosuppressive drugs, FK506 and rapamycin, played a critical role in the demonstration that the inhibition of the PPIase activity of the immunophilins is not responsible for the immunosuppressive effects of their respective ligands. The structures of CsA, FK506 and rapamycin are shown in Fig. 1. The structural differences between these two classes of immunosuppressive agents explains their recognition by two different classes of immunophilins, cyclophilins and FKBPs and their ability to inhibit the PPIase activity of their binding proteins.

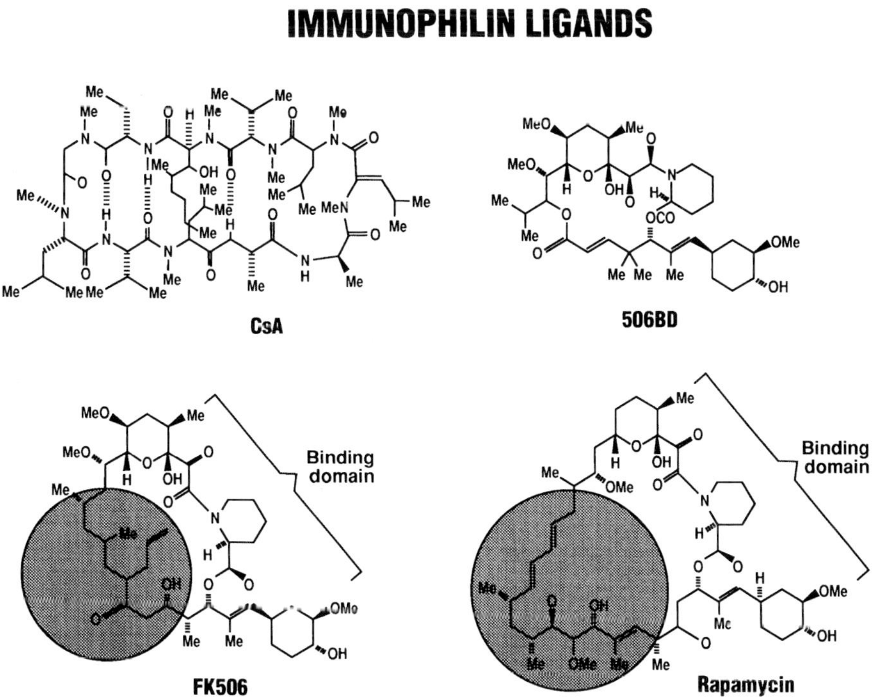

Fig. 1. Structures of Immunophilin ligands. (reproduced with permission from Ref. 9b)

All three compounds are strong inhibitors of the PPIase activities of their binding proteins. A structural motif, shared by FK506 and rapamycin (binding domain) was predicted to be recognized by their common binding protein, FKBP [10]. CsA, which interacts with a different immunophilin, lacks this motif. NMR and crystallographic studies of FKBP and CyPA and of their respective complexes with FK506 and CsA definitively identified the catalytic domains of FKBP and CyPA and the binding motifs of the three ligands illustrated in Fig. 1 [11–15]. The crystal structures of the FKBP12-FK506 complex and the solution structure of the CyPA-CsA complex are illustrated in Fig. 2A and 2B. Although CyPA and FKBP have different architectures they both have two protruding loops that form the sides of a shallow cleft forming the binding sites for CsA and FK506 acting as a transition state analog of the peptide substrates and thereby explaining their potent inhibition of the PPIase activity [16, 17].

Although both FK506 and rapamycin inhibit PPIase with similar K_i values (0.4 nM for FK506 and 0.2 nM for rapamycin) they were shown to inhibit different steps in T cell activation [18]. Like CsA, FK506 inhibits the Ca^{2+}-dependent expression of early genes [such as interleukin-2 (IL-2)] in response to the occupancy of T-cell receptors, whereas rapamycin interferes with a Ca^{2+}-independent pathway leading to cell differentiation and proliferation in response to the occupancy of IL-2 receptors. Thus, the inhibition of PPIase activity could not explain the different biological responses elicited by these two compounds. This discrepancy combined

(a)

(b)

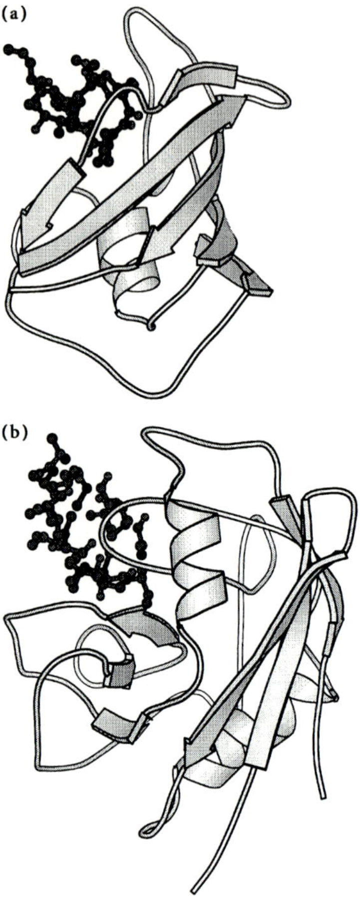

Fig. 2. (a) Ribbon representation of the crystal structure of FKBP12-FK506. FK506 is shown in ball and stick representation (PDB code 1TCO [31b]). **(b)** Solution structure of CyPA-CsA. CsA is shown in ball and stick representation (PDB code 3CYS [78]). The ribbon diagrams were generated with MOLSCRIPT [79]

with the structural differences shown in Fig. 1 led Schreiber and colleagues [10] to postulate the existence of an "effector domain" (outlined by a shaded circle in Fig. 1) to be recognized by different cellular targets responsible for the different cellular responses. This effector domain is shown in Fig. 2A and 2B as a protruding motif in the structures of FKBP12-FK506 and CyPA-CsA . To add support to this model, an analog of FK506, 506DB, lacking the putative effector domain, was synthesized and shown to inhibit the PPIase activity of FKBP but, as predicted by the proposed model, 506DB lacked the immunosuppressive activities of both FK506 or rapamycin [18]. Inhibition of the immunosuppressive effects of FK506 and rapamycin by 506DB could be explained by the dissociation of FKBP-FK506 and FKBP-rapamycin complexes induced by 506DB suggesting that the complexes were the functional entities rather than the drugs themselves. These observations strongly supported the mechanism based on a "gain of function" rather than the original "loss of function" mechanism based on PPIase inhibition and prompted the search of the target of the effector domains [10, 18, 19].

3
Identification of the Target of FK506/FKBP12 and CyPA-CsA

3.1
Ca^{2+}-Dependent Binding of Calcineurin to FKBP12-FK506 and CyPA-CsA

The now widely used method to detect protein-protein interactions based on the ligand-dependent binding of a protein to another protein fused to glutathione-S-transferase (GST) was ideally suited for the isolation of the target protein(s) recognized by FK506 and CsA [20]. Using this method Liu et al. [21] demonstrated that the same protein bound to the fusion proteins GST-FKBP12 and GST-CyPA expressed in *E. coli* but only in the presence of their respective ligands, FK506 and CsA and not to GST-FKBP12 in the presence of rapamycin. The displacement of the protein bound to FKBP-FK506, but not to CypA-CsA, by rapamycin and 506DB was consistent with the identification of the FKBP-FK506 complex as the active entity [10]. Conversely CsA did not dissociate the protein from FKBP12-FK506. This protein was identified as calcineurin. Calcineurin (also called protein phosphatase-2B), the only protein phosphatase under the control of Ca^{2+} and calmodulin, is a heterodimer of a 59 kDa catalytic subunit tightly bound to a Ca^{2+}-binding protein of the "EF-hand" family, calcineurin B [22]. The two subunits of calcineurin bound to the fusion proteins were identified by their cross reactivity with anti-calcineurin antibodies and Ca^{2+} binding to calcineurin B [21]. Consistent with the observation that both FK506 and CsA block the same Ca^{2+}-dependent step in T cell activation, the binding of calcineurin to GST-FKBP-FK506 and GST-CypA-CsA was Ca^{2+} dependent. Also consistent with the proposed model, calcineurin did not bind to the FKBP-rapamycin complex later shown to interfere with a distinct Ca^{2+}-independent pathway in T cell activation [23].

3.2
Inhibition of Calcineurin Activity by FKBP-FK506 and CypA-CsA

A critical test was to demonstrate that FKBP12-FK506 and Cyp-CsA inhibited the phosphatase activity of calcineurin. As summarized in Table 1 both FKBP12-FK506 and CypA-CsA inhibit the Ser/Thr phosphatase activity of calcineurin with K_i values in the nM range and the inhibition requires the presence of both the binding proteins and their ligands [21, 24].

Table 1. Specific inhibition of calcineurin by immunophilin ligands

Inhibitor	PPase-1	PPase-2A	PPase-2C	Calcineurin		References
				Cn 61	CN 45	
			K_i (nM)			
FKBP-FK506	>1000	>1000	>1000	30	<50	24
CyP-CsA	>1000	>1000	>1000	40	ND	21
Okadaic acid	10.0	0.1	>10,000	>3000	2500	25a
Microcystin	<0.1	<0.1	>4000	200	500	25b
Calyculin	2	1		1500	1100	25c

The inhibition by FKBP-FK506 is slowly reversed upon dilution or the addition of high concentrations of rapamycin. Failure to inhibit the three other neutral protein phosphatases, PP-1, PP-2A and PP-2C at μM concentrations of FKBP12-FK506 and CypA-CsA firmly established the high specificity of these inhibitors for calcineurin as opposed to other protein phosphatases thereby providing the long-awaited specific inhibitors needed to identify the physiological roles of calcineurin, up to then a phosphatase in search of function(s) [25].

3.3 Correlation Between Calcineurin Inhibition and Immunosuppressive activity

A further test to definitively identify calcineurin as the target of the immunosuppressive drugs was to correlate their ability to inhibit calcineurin with their immunosuppressive activity. As illustrated in Table 2, in contrast to the previously reported lack of correlation between the immunosuppressive activities of CsA and FK506 derivatives with their ability to inhibit the PPIase activity, there is an excellent correlation between their immunosuppressive activities and their ability to inhibit calcineurin [24].

The membrane permeability of FK506, rapamycin, and CsA and the ubiquitous distribution of their binding proteins made it possible to study the immunosuppressive effects of the drugs in vivo [26]. In intact cells overexpression of calcineurin was shown to render the cells resistant to the inhibitory effects of the two drugs and to increase the induction of a reporter gene under the control of the IL-2 promoter mediated by the transcription factor NFAT (Nuclear Factor of Activated T cells) [27]. Rapamycin had no effect on the induction of the same reporter gene. This experimental approach, later extended to the use of transgenic animals overexpressing a constitutively active form of calcineurin, has now been widely used to study the role of calcineurin *in vivo*. Equally successful was the combined use of these specific inhibitors of calcineurin and of yeast genetics to establish the general role of calcineurin in the regulation of gene expression [28].

Table 2. Correlation between immunosuppressive activities and inhibition of calcineurin for FK506 and CSA analogs[a]

Compounds	K_i (nM)		IC_{50} (nM) NFAT activity[b]	K_i (nM) calcineurin
	CyPA	FKBP12		
CSA	6		3.5	40
MeAla^6-CSA	9		3200	>1000
FK506		1.0	0.5	34
15-0-DeMe-FK250		15.0	>800	1600

[a] from Ref. [24].
[b] measured by NFAT-driven β-galactosidase activity.

4 Structure of the Calcineurin-FKBP12-FK506 Complex

One question remained to be answered: how two structurally different but functionally similar effector molecules, FKBP12-FK506 and CypA-CsA, bind competitively to the same site on calcineurin and exert the same inhibitory effect on its activity. The answer to this question depended on the elucidation of the structure of calcineurin bound to the immunophillin-drug complexes.

4.1 Structure of Calcineurin

The functional domain organization of the catalytic subunit of calcineurin is conserved from yeast to man [22]. First revealed by limited proteolysis [29] it was further refined by the use of synthetic peptides, site directed mutagenesis, and NMR studies [30] and confirmed by the elucidation of the crystal structures of recombinant alpha isoform of human calcineurin and of the truncated form of bovine brain calcineurin [31].

As schematized in Fig. 3, the amino-terminal two-thirds of the molecule, resistant to proteolysis, contains the catalytic domain and the calcineurin B-binding

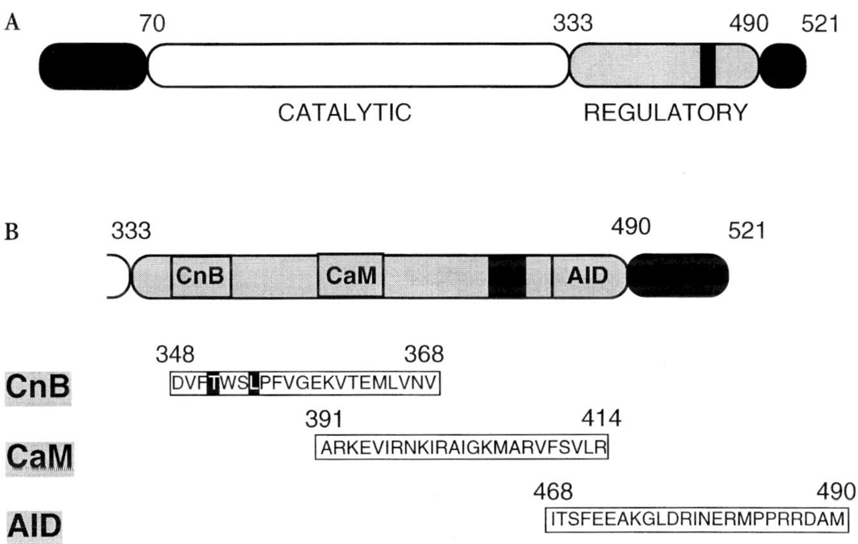

Fig. 3. Functional domain organization of calcineurin A. **A** Schematic representation of the three mammalian isoforms of calcineurin A. The variable regions and a 10-amino acid insert, resulting from alternative splicing in the α and β isoforms, are shown in *black*. **B** Extended representation of the regulatory domain and the calcineurin B-binding domain (CnB). The calmodulin-binding (CaM) and autoinhibitory domain (AID) are *boxed*. The numbering of the amino acids is that of calcineurin Aα. Residues critical for interaction with Cyp and FKBP12 are shown by *white letters on a black background*. (reproduced from [22c])

domain. The carboxyl-terminal third of the molecule, readily cleaved by proteases, contains the calmodulin-binding domain and an autoinhibitory domain [29]. In the crystal structure of the full length protein [31a] the catalytic center is blocked by two short α-helices formed by the eighteen residue peptide previously identified as an inhibitory domain [32]. The calmodulin-binding domain is not visible in the electron density map suggesting that this domain has a disordered structure consistent with its extreme sensitivity to proteolytic attack. A truncated form of calcineurin, lacking these two regulatory domains, no longer binds or is dependent on calmodulin for activity [29]. It is often referred to as the constitutive or Ca^{2+}-independent form of the enzyme that remains sensitive to FK506 and CsA. Its apparent Ca^{2+} independence stems from the fact that it has a very high affinity for Ca^{2+} and still depends on Ca^{2+} binding to calcineurin B for activity [33]. Thus, the phosphatase activity of calcineurin is under the control of two structurally similar but functionally distinct Ca^{2+}-regulated proteins. The interaction of calcineurin B with calcineurin A is apparently mediated by the very high affinity Ca^{2+} site of calcineurin B [34] while Ca^{2+} binding to the moderate affinity sites is required for activation and calmodulin activation by a yet unknown mechanism [33]. Ca^{2+}-binding to calmodulin is required for its interaction with calcineurin and the full activation of its phosphatase activity [22]. The highly cooperative calmodulin-mediated and Ca^{2+}-dependent activation of calcineurin allows the enzyme to respond to very narrow Ca^{2+} thresholds.

The structure of the catalytic and calcineurin B-binding domains of the truncated derivative of calcineurin complexed with FKBP12-FK506 (Fig. 4) is essentially identical to that of the same domains in the full length protein [31b]. The catalytic domain consists of a sandwich of two sheets of β strands flanked on one side by three α-helices and three β-strands and on the other by an all helical structure. The catalytic center is defined by its two metal ions, iron and zinc. The last β-sheet extends into a five-turn amphipathic α-helix that constitutes the calcineurin B-binding domain of calcineurin A. The top face of this helix is covered by a groove formed by the carboxyl- and amino-terminal lobes of calcineurin B together with the carboxyl terminal strand of calcineurin B.

The structure of calcineurin is consistent with the proposed mechanism for the calmodulin activation of calcineurin and explains its activation upon removal of the calmodulin-binding domain and a putative inhibitory domain by limited proteolysis [29]. According to this mechanism, binding of calmodulin to the calmodulin-binding domain results in the displacement of the autoinhibitory domain and exposure of the active site. The flexible calmodulin-binding domain, readily accessible to calmodulin and proteases, is also consistent with this mechanism. The exposure of the active site upon calmodulin binding also explains the calmodulin-dependent inactivation of calcineurin resulting from the oxidation of Fe^{2+} in the active site by superoxide anion [35]. The structures of the Ca^{2+}-saturated forms of calcineurin do not provide any information on the requirement of calcineurin B for the phosphatase activity of calcineurin. The phosphatase activity of calcineurin A is usually undetectable or very low [36] and the deletion of calcineurin B gene in yeast is sufficient to yield a calcineurin null phenotype [37]. The requirement for Ca^{2+} binding to the low affinity sites of calcineurin B for its calmodulin activation is also not yet understood [33].

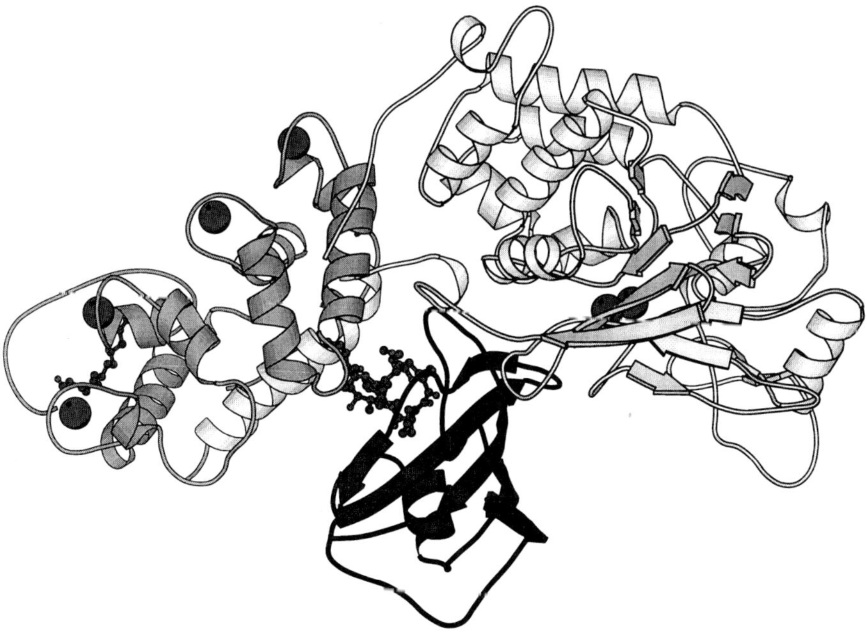

Fig. 4. Ribbon representation of the crystal structure of truncated calcineurin complexed with FKBP12-FK506. Calcineurin A is shown in *light gray*, calcineurin B in *dark gray* and FKBP12, the four Ca^{2+} in the calcineurin B sites and Fe^{3+} and Zn^{2+} in the active site in *black*. FK506 is shown in ball and stick representation. The ribbon diagram was generated with MOLSCRIPT [79]. (PDB code 1TCO [31b])

4.2 Binding Domain of Calcineurin for FKBP12-FK506

The anchoring site of the two immunosuppressive drugs, CsA and FK506, is composed of calcineurin B and the calcineurin B-binding domain of calcineurin A as predicted on the basis of site directed mutagenesis and crosslinking experiments [38]. It illustrates the essential role of calcineurin B in the interaction between the drugs and calcineurin first revealed by crosslinking experiments [39] and provides a molecular basis for the exquisite specificity of the immunosuppressive agents for calcineurin as opposed to other protein phosphatases [24]. It also explains the requirement of calcineurin B for the inhibition of calcineurin by CyPA-CsA and FKBP12-FK506 [36b]. The interface of the calcineurin B-binding domain and calcineurin A forms the binding site of FK506 while the more polar bottom face of the calcineurin B-binding helix together with calcineurin B and the catalytic domain of calcineurin A makes contact with distinct regions of FKBP12. Two thirds of the surface contact between FKBP12-FK506 and calcineurin B comes from the latch region identified by Milan et al. [40] as the major site of interaction of calcineurin B with the CypA-CsA complex. The recognition of this latch region, formed by calcineurin B upon binding to calcineurin A, by each of the two immunosuppressive complexes, explains their competitive binding to calcineurin [21], and the calcineurin A requirement for CsA/calcineurin B interaction [39]. Yeast calcineurin A mutants that are specifically resistant to FK506 or cyclosporin indicated that the interaction sites are overlapping but not identical [41].

The structure of calcineurin complexed with FKBP12-FK506 is consistent with the non competitive inhibition of calcineurin by the CypA-CsA complex [42] but does not explain the stimulation of its *p*-nitrophenyl phosphatase activity [21, 36b]. In the Ca^{2+} saturated forms of calcineurin the binding domain for the immunophilin-drug complexes are readily accessible and calmodulin stimulates but is not required for interaction. The structure of the Ca^{2+}-depleted calcineurin is needed to explain why Ca^{2+} binding to calcineurin B is needed to identify the nature of the conformational change accompanying Ca^{2+} binding to calcineurin B that is critical for the binding of the FKBP12-FK506 complex to the full length protein [21].

5 Endogenous Analogs of FK506

IP_3 and ryanodine receptors interact with and are regulated by FKBP12 [43, 44]. The findings that calcineurin is bound to the IP_3 receptor-FKBP complex in a Ca^{2+}-dependent fashion as it is to the FK506-FKBP complex, and is released from the receptor by FK506 and rapamycin led to the proposal that a FK506-like motif on the receptor acts as an endogenous analog of FK506 [44b]. The identification of a Leu-Pro dipeptide (residues 1400–1401) as the moiety recognized by FKBP12 because of the previously reported structural similarity of the drugs to the transition state structures of *cis-trans* leucyl-prolyl and valyl-prolyl bonds, preferred substrates of FKBP12 among succinyl-Ala-X-Pro-Phe synthetic peptides, added support to this hypothesis [45]. In this case the interaction does not result in inhibition but serves as way to bring the substrate close to the receptor to allow the regulation of its activity by a dephosphorylation mechanism [44b]. The L/VP dipeptide is conserved in motifs similar to the ^{1396}CNSL**LP**LDDIV motif of the IP_3 receptor type I, present in the IP_3 receptors types II and III, the ryanodine receptor type I, and the transforming factor TGFβ type I, but is absent in the TGFβ receptor type II, which does not bind FKBP12 [43a, 44a, 45a, 46]. The PPIase activity of FKBP12 is not required for binding or activity since mutants of FKBP12 lacking rotamase activity interact with and regulate the channel activity of the receptors [45a, 47]. The rapamycin inhibition of the FKBP12-mediated regulation of the receptor, originally attributed to the inhibition of the PPIase activity of FKBP12, is the result of the dissociation of calcineurin from the receptor upon rapamycin binding to FKBP12 [44b]. Thus, the calcineurin modulation of these channels, mediated by FKBP12 targeting of calcineurin to these receptors, cannot be identified by the usual differential inhibitory effect of FKBP12 and rapamycin. It remains to be shown that rapamycin-induced dissociation of the FKBP12-calcineurin complex, which does not interfere with calcineurin activity, is sufficient to inhibit the modulation of these channels. The calcineurin-catalyzed dephosphorylation of the IP_3 receptor sites implies that, unlike FKBP12-FK506, the endogenous receptor-FKBP complex acts as a stimulator rather than an inhibitor of calcineurin-mediated dephosphorylation.

As described above for the Ca^{2+} channel receptors, HIV-1 capsid protein and rhodopsin, acting as natural analogs of CsA, form CsA-sensitive complexes with cyclophilin believed to be recognized by yet unidentified target proteins. The interaction of CypA with the HIV-1 capsid protein is required for viral infectivity and the interaction of Drosophila NinaA with rhodopsin is needed for the recycling of the photoreceptor [48].

6
Physiological Roles of Calcineurin

6.1
T Cell Activation

The specific inhibition of calcineurin by the immunosuppressive drugs opened the way to elucidate a signal transduction pathway from the plasma membrane to the nucleus. Taking advantage of the different electrophoretic mobilities of phosphorylated proteins McCaffrey et al. [49] isolated a protein whose dephosphorylation was inhibited by FK506 but not by rapamycin. The identification of this protein, a transcription factor of the NFAT family, as the substrate of calcineurin responsible for the induction of cytokines upon T cell activation was the last step needed to link T cell receptor activation to the regulation of gene expression.

According to the mechanism illustrated in Fig. 5 the increase in intracellular Ca^{2+} concentration following the occupancy of the T cell receptor results in calcineurin activation, specific dephosphorylation of many phosphorylated sites located in the regulatory domain of NFAT, and translocation of NFAT to the nucleus [50]. Since the interaction of NFAT with calcineurin is not dependent on the phosphorylation state of NFAT but on the presence of Ca^{2+}, calcineurin remains bound to NFAT and is translocated to the nucleus as a complex with NFAT [51]. A prolonged exposure to Ca^{2+} and Ca^{2+} entry to refill the Ca^{2+} stores is needed to insure the sustained interaction of NFAT with calcineurin whose activation is apparently required to enhance the DNA binding and transcriptional activity of NFAT [52]. As illustrated in Fig. 5, together with transcription factors induced by the Ca^{2+} stimulation of protein kinase C, NFAT is responsible for the induction of many cytokines including that of atleast one of the NFAT isoforms, NFAT2 [53].

The inhibitory effects of the immunosuppressive drugs, FK506 and CsA on the nuclear translocation of NFAT have played a major role in establishing this Ca^{2+}-dependent, calcineurin-mediated, transduction to the nucleus of an external signal acting on the Tcell receptor [52b, 54]. Equally useful was the ability to overexpress calcineurin and NFAT fused with the fluorescent green protein (GFP) in order to monitor their FK506-sensitive translocation from the cytoplasm to the nucleus following T cell activation [55]. The calcineurin-dependent translocation of NFAT to the nucleus is readily reversible upon treatment with FK506 or CsA, depletion of the Ca^{2+} stores, or removal of the external signal that allows phosphorylation of NFAT and its export to the cytoplasm [54a, 54b, 55]. Several kinases, glycogen synthase kinase-3, casein kinase 1, MEKK1 and the Jun *N*-terminal kinase have been implicated in this process whose precise mechanism is not yet fully understood [50d].

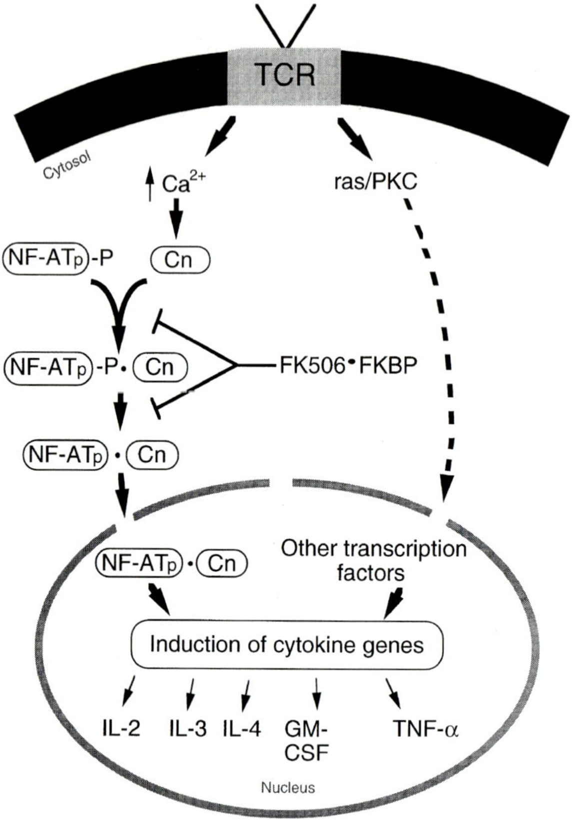

Fig. 5. Schematic representation of calcineurin mediated T cell activation as described in Ref. 50d. Stimulation of the T cell receptor (TCR) is accompanied by a sustained increase in intracellular Ca^{2+}, activation of calcineurin (Cn), interaction of calcineurin with and dephosphorylation of phosphorylated cytoplasmic NFATp and translocation of dephosphorylated NFATp complexed with calcineurin to the nucleus and induction of the cytokine genes [interleukin (IL) -2, -3, -4, granulocyte-macrophage colony-stimulating factor (GM-CFS), and tumor necrosis factor (TNFα). Gene expression may require the concomitant activation of the protein kinase C (PKC)/ras21 pathway

The list of calcineurin-sensitive transcription factors keeps increasing. Some, like Elk-1, are substrates of calcineurin [56]. The activity of many others, NFκB, Jun/ATF2, CREB, Nur77 and MEF2C is indirectly modulated by calcineurin regulating the expression or activity of other factors [50d].

6.2 Ion Transport and Cell Integrity in Yeast

The key role of calcineurin in the regulation of gene expression is now well established. In the budding yeast, *Saccharomyces cerevisiae*, calcineurin regulates the expression of many genes involved in ion transport and cell integrity by a mechanism similar to that described for T cell activation [22d, 28, 50e]. The stimulation of calcineurin following an increase of intracellular concentration also results in the dephosphorylation and nuclear translocation of a transcription factor, TCN1p/CRZ1p, that is responsible for the induction of genes involved in Na^+, Ca^+, H^+ transport as well as the gene of the glucan synthase complex, *GSK2*, that is required to maintain cell viability following exposure to pheromones [57].

6.3 Regulation of Neuronal Functions

The diversity of calcineurin functions is particularly evident in brain where the knock-out of the neural isoform of calcineurin is accompanied by pronounced neurological disorders [58]. The immunosuppressive drugs together with the cell specific overexpression of the constitutive form of calcineurin have helped to identify several of these calcineurin-mediated processes [59]. A major role of calcineurin in the brain is to trigger a protein phosphatase cascade by dephosphorylating DARPP32 and inhibitor-1, early identified as good substrates of calcineurin [60]. These two proteins are inhibitors of protein phosphatase-1 when phosphorylated by cAMP-dependent protein kinase. Their Ca^{2+}-dependent dephosphorylation by calcineurin and the resulting activation of protein phosphatase-1 provides a general mechanism to explain the antagonistic effects of Ca^{2+} and cAMP in the regulation of many neuronal processes [61]. Accordingly, this Ca^{2+}-dependent activation of protein dephosphorylation is inhibited by FK506 and CsA as well as by the inhibitors of protein phosphatase-1, okadaic acid, microcystin and calyculin. The calcineurin-mediated inhibition of the excitability of striatal neurons upon binding of glutamate to the NMDA (*N*-methyl-D-aspartate) receptors is such an example [62]. Another example is the inhibition of the conversion of early to late long term potentiation, and loss of long term memory in transgenic mice overexpressing the constitutive form of calcineurin in hippocampal neurons [63]. The Ca^{2+}-mediated down regulation of the expression of genes involved in synaptic plasticity which are under the positive control of the cAMP-response element binding protein, CREB, also depends on a protein phosphatase cascade triggered by calcineurin [64].

In hippocampal neurons CREB phosphorylation by calmodulin kinase IV, depends on the inactivation of calcineurin upon prolonged exposure to Ca^{2+} [65]. A Ca^{2+}/calmodulin-dependent inactivation of calcineurin resulting from the oxidation of the active site Fe^{2+} by superoxide, was proposed as a mechanism to regulate calcineurin activity in response to oxidative stress in yeast cells [35]. The activity-dependent production of reactive oxygen species, the inactivation of calcineurin following prolonged Ca^{2+} signals, and its protection by scavengers of the superoxide in hippocampal neurons is the first evidence that such a mechanism operates in mammalian cells [65].

7 Other Biological Functions of Calcineurin

Calcineurin has been implicated not only in the regulation of immune responses but in many other biological processes. NFAT is a member of a family of related proteins, originally believed to be specific for T cells but which have recently be shown to be broadly distributed, extending the potential sites of action of calcineurin to the regulation of gene expression in many tissues [50 and ref. therein]. Cardiac hypertrophy and congestive heart failure, which could be prevented by treatment with CsA, have been reported in transgenic mice overexpressing the constitutive form of calcineurin [66]. It was proposed that slow muscle fiber differentiation induced by chronic stimulation is the result of a calcineurin-mediated dephosphorylation of the NFAT isoform, NFAT3 [67]. The protective effect of FK506 and CsA against glutamate neurotoxicity linked to neuronal cell death has been attributed to the inhibition of calcineurin activity preventing the dephosphorylation and activation of NO synthase [68]. A similar neuroprotective effect of FK506 in an *in vivo* model of focal cerebral ischemia suggests the clinical potential of FK506 for the treatment of stroke [69].

8 Perspectives

The identification of calcineurin as the target of the CyPA-CsA and FKBP12-FK506, less than ten years ago, followed by the identification of the transcription factor NFAT as a specific substrate of calcineurin revealed a novel Ca^{2+}-mediated signal transduction pathway conserved from yeast to man. During the past five years the complexity of this pathway has become evident with the identification of many immunophilin isoforms, the broad tissue distribution of NFAT and the myriad of genes under its control. While the diversity of processes affected by these fascinating drugs revealed alternative potential applications they also illustrate the difficulty in designing drugs devoid of side effects.

CsA and FK506 are potent and specific inhibitors of all known isoforms of calcineurin and not other protein phosphatases *in vitro* but can interfere with the activity of the latter *in vivo*. There is also recent evidence that the inhibition of the PPIase activity of the immunophilins could be responsible for some side effects. There is suggestive evidence that the inhibition of the PPIase of FKBP and CyP is implicated in neuronal cell death in familial amyotrophic lateral sclerosis [70] the regulation of Ca^{2+}-dependent K+ channels [71] and memory formation [72]. Derivatives of FK520 (ascomycin) have already been synthesized that have a low affinity for FKBP12 in the absence of calcineurin but binds more tightly to FKBP12 in the presence of calcineurin and are therefore better immunosuppressants than PPIase inhibitors [73].

Another problem encountered in analyzing the effects of these drugs *in vivo* is the dependence of their effects on the concentrations of their binding proteins. In any given cell the potency and specificity of CsA and FKBP depends on the concentration of immunophilins as well as that of their target, calcineurin, and its substrates. Thus, experiments based on overexpression of any one of these proteins *in vivo* or in intact cells should be interpreted with caution. Also to bear in mind is the existence of many isoforms of CyP and FKBP. Many of these isoforms do not interact with calcineurin but with other targets which are either inhibited or potentiated by CsA and FK506. For example FK506 binding to FKBP59 and CsA binding to Cyp40 potentiate glucocorticoid receptor-mediated gene expression by a calcineurin-independent mechanism [74] whereas binding of FKBP12 to the TGFβ receptor prevents its phosphorylation and activation [75] and binding of FK506 to FKBP59 has no effect on its interaction with a newly identified target, phytanoyl-COA-α-hydroxylase involved in the Refsun disease [76].

The continuously expanding list of calcineurin functions and the myriad of transcription factors under the control of its major substrate NFAT emphasizes the need for agents acting on specific substrates. Preliminary evidence that different substrates interact with different domains of calcineurin suggests that such compounds can be designed [77]. The design of such drugs will depend on the definitive demonstration that different substrates recognize different sites on calcineurin.

References

1. Borel JF (1989) Pharmacol Rev 41:251
2. Handschumaker RE, Harding MW, Rice J, Drugge RJ, Speicher DW (1984) Science 226:544
3. Takahashi N, Hayano T, Suzuki M (1989) Nature 337:473
4. Fischer G, Wittmann-Liebold B, Lang K, Kielhaber T, Schmid FX (1989) Nature 337:476
5. Tanaka H, Kuroda A, Marusawa H, Hatanaka H, Kino T, Hoto T, Hashimoto M (1987) J Am Chem Soc 109:5031
6. Findlay AJ, Radics L (1980) Can J Chem 58:579
7. Siekerska JJ, Hung SHY, Poe M, Lin CS, Sigal NH (1989) Nature 341:755
8. Harding MW, Galat A, Uehling DE, Schreiber SL (1989) Nature 341:758
9. For reviews see a) Rosen MK, Schreiber SL (1992) Angew Chem Int Ed Engl 31:384 b) Schreiber SL, Liu J, Albers MW, Rosen MK, Standaert RF, Wandless TJ, Somers PK (1992) Tetrahedron 448:2545 c) Schreiber SL, Crabtree GR (1992) Immunology Today 13:136 d) Braun W, Kallen J, Mikol V, Walkinshaw MD, Wuthrich K (1995) FASEB J 9:63
10. Schreiber SL (1991) Science 251:283
11. Moore JA, Peattie DA, Fitzgibbon MJ, Thompson JA (1991) Nature 351:248
12. Michnik SW, Rosen MK, Wandless TJ, Karplus PA, Schreiber SL (1991) 251:836
13. Van Duyne GD, Standaert RF, Karplus PA, Schreiber SL, Clardy J (1993) J Mol Biol 229:105
14. a) Ke H, Zydowsky LD, Liu J, Walsh CT (1991) Proc Natl Acad Sci USA 88:9483
b) Mikol V, Kallen J, Walkinshaw MD (1994) Proc Natl Acad Sci USA 91:5183
15. Ottiger M, Zerbe O, Guntert P, Wuthrich K (1997) J Mol Biol 272:64
16. Albers MW, Walsh CT, Schreiber SL (1990) J Org Chem 55:4984
17. Kofron JL, Kusmic P, Kishore V, Colon-Bonilla E, Rich DE (1991) Biochemistry 30:6127
18. a) Bierer BE, Somers PK, Wandless TJ, Burakoff SJ, Schreiber SL (1990) Science 250:556
b) Somers PK, Wandless TJ, Schreiber SL (1991) J Am Chem Soc 113:8045
19. Bierer BE, Mattila PS, Standaert RF, Herzenberg LA, Burakoff SJ, Crabtree GR, Schreiber SL (1990) Proc Natl Acad Sci USA 87:9231
20. Friedman J, Weisman I (1991) Cell 66:799
21. Liu J, Farmer Jr. JD, Friedman J, Weissman I, Schreiber SL (1991) Cell 66:807
22. For reviews on calcineurin see: a) Klee CB, Draetta GF, Hubbard MJ (1988) Adv Enzymol Relat Areas Mol Biol 61:149 b) Kincaid RL (1993) Adv Second Messenger Phosphoprotein Res 27:1 c) Klee CB, Ren H, Wang X (1998) J Biol Chem 273:13367 d) Klee CB, Wang X, Ren H (1999) Calcium-regulated protein dephosphorylation. In: Carafoli E, Klee C (eds) Calcium as a cellular regulator. Oxford University press. New York, p 344 e) Hemenway CS, Heitman J (1999) Cell Biochem Biophys 30:115
23. Brown EJ, Schreiber SL (1996) Cell 86:517
24. Liu J, Albers MW, Wandless TJ, Luan S, Alberg DG, Belshaw PJ, Cohen P, MacKintosh C, Klee CB, Schreiber SL (1992) Biochemistry 31:3896
25. a) Bialojan C, Takai A (1988) Biochem J 256:283 b) Cohen P, Holmes CF, Tsukitani Y (1990) Trends Biochem Sci 15:98 c) Ishihara H, Martin BL, Brautigan DL et al. (1989) Biochem Biophys Res Commun 159:871

26. Fruman DA, Klee CB, Bierer BE, Burakoff SJ (1992) Proc Natl Acad Sci USA 89:3686
27. a) O' Keefe SJ, Tamura J, Kincaid RL, Tocci MJ, O'Neill EA (1992) Nature 357:692 b) Clipstone NA, Crabtree GR (1992) Nature 357:695
28. Cardenas ME, Lorenz M, Hemenway C, Heitman J (1994) Perspect Drug Discovery Design 2:103
29. Hubbard MJ, Klee CB (1989) Biochemistry 28:1868
30. a) Watanabe Y, Perrino BA, Chang BH, Soderling TR (1995) J Biol Chem 270:456 b) Sikkink R, Haddy A, MacKelvie S, Mertz P, Litwiller R, Rusnak F (1995) Biochemistry 34:8348 c) Roksz LL, O'Keefe SJ, Parsons JN, Cameron PM, Burbaum JJ (1995) Protein Expr Purif 6:655 d) Perrino BA, Ng LY, Soderling TR (1995) J Biol Chem 270:7012 e) Anglister J, Ren H, Klee CB, Bax A (1995) FEBS Lett 375:108
31. a) Kissinger CR, Parge HE, Knighton DR, Lewis CT, Pelletier LA, Tempczyk A, Kalish VJ, Tucker KD, Showalter RE, Moomaw EW, Gastinel LN, Habuka N, Chen X, Maldonado F, Barker JE, Bacquet R, Villafranca JE (1995) Nature 378:641 b) Griffith JP, Kim JL, Kim EE, Sintchak MD, Thompson JA, Fitzgibbon MJ, Fleming MA, Caron PR, Hsiao K, Navia MA (1995) Cell 82:507
32. Hashimoto Y, Perrino BA, Soderling TR (1990) J Biol Chem 265:1924
33. Stemmer PM, Klee CB (1994) Biochemistry 33:6859
34. Kakalis LT, Kennedy M, Sikkink R, Rusnak F, Armitage IM (1995) FEBS Lett 362:55
35. Wang X, Culotta VC, Klee CB (1996) Nature 383:434
36. Merat DL, Hu ZY et al. (1985) J Biol Chem 260:11053 b) Haddy A, Swanson SK, Born TL, Rusnak F (1992) FEBS Lett 314:37
37. Cyert MS, Thorner J (1992) Proc Natl Acad Sci USA 121:3460
38. a) Clipstone NA, Fiorentino DF, Crabtree GR (1994) J Biol Chem 269:26431 b) Husi H, Luyten MA, Zurini MG (1994) J Biol Chem 269:14199
39. a) Li W, Handschumacher RE (1993) J Biol Chem 268:14040 b) Ryffel B, Woerly G, Murray M, Eugster HP, Car B (1993) Biochem Biophys Res Commun 194:1074
40. Milan D, Griffith J, Su M, Price ER, McKeon F (1994) Cell 79:437
41. a) Cardenas ME, Muir RS, Breuder T, Heitman J (1995) EMBO J 14:2772 b) Kawamura A, Su MS (1995) J Biol Chem 270:15463
42. Etzkorn FA, Chang ZY, Stolz LA, Walsh CT (1994) Biochemistry 33:2380
43. a) Jarayaman T, Brillantes AM, Timerman AP, Erdjument-Bromage H, Fleischer S, Tempst P, Marks AR (1992) J Biol Chem 267:9474 b) Brillantes AB, Ondrias K, Scott A, Kobrinski E, Ondriasova E, Moschella MC, Jaramayan T, Landers M, Ehrlich BE, Marks AR (1994) Cell 77:513 b)Timerman AP, Ogunbumni E, Freund E, Wiederrecht G, Marks AR, Fleischer S (1993) J Biol Chem 268:22992
44. a) Cameron AM, Steiner JP, Sabatini DM, Kaplin AI, Walensky LD, Snyder SH (1995) Proc Natl Acad Sci USA 92:1784 b) Cameron AM, Steiner JP, Roskams AJ, Ali SM, Ronnett GV, Snyder SH (1995) Cell 83:463
45. a) Cameron AM, Nucifora FC Jr, Fung ET, Livingston DG, Aldape RA, Ross CA, Snyder SH (1997) J Biol Chem 272:27582 b) Albers MW, Walsh CT, Schreiber SL (1990) J Org Chem 55:4984
46. a) Derynck R (1994) Trends Biochem Sci 19:548 b) Wang T, Donahoe PK, Zervos AS (1994) Science 265:674 c) Wang T, Li B-Y, Danielson PD, Shah PC, Rockwell S, Lechleider RJ, Martin J, Manganaro T, Donahoe PC (1996) Cell 86:435
47. Timerman AP, Wiederrecht G, Marcy A, Fleischer S (1995) J Biol Chem 270:2451
48. For a review see: Luban J (1996) Cell 87:1157
49. McCaffrey PG, Perrino BA (1993) J Bio Chem 269:30445
50. For reviews see: a) Rao A, Luo C, Hogan PG (1997) Annu Rev Immunol 15:707 b) Crabtree GR, Clipstone NA (1994) Annu Rev Biochem 63:1045 c) Crabtree GR (1999) Cell 96:611 d) López-Rodríguez C, Aramburu J, Rakeman AS, Copeland NG, Gilbert DJ, Thomas S, Disteche C, Jenkins NA, Rao A (1999) Cold Spring Harb Symp Quant Biol In press e) Aramburu J, Rao A, Klee CB (2000) Curr Top Cell Reg 36:237
51. a) Wesselborg S, Fruman DA, Sagoo JK, Bierer BE, Burakoff SG (1996) J Biol Chem 271:1274 b) Loh C, Carew JA, Kim J, Hogan PG, Rao A (1996) Mol Cell Biol 16:3945 c) Luo C, Shaw KT, Raghavan A, Aramburu J, Garcia-Cozar F, Perrino BA, Hogan PG, Rao A (1996) Proc Natl Acad Sci USA 93:8907 d) Masuda ES, Liu J, Imamura R, Imai SI, Arai KI, Arai N (1997) Mol Cell Biol 17:2066

52. a) Park J, Yaseen NR, Hogan PG, Rao A, Sharma S (1995) J Biol Chem 270:20653 b) Shaw KT, Ho AM, Raghavan A, Kim J, Jain J, Park J, Sharma S, Rao A, Hogan PG (1995) Proc Natl Acad Sci USA 92:11205
53. a) Northrop JP, Ho SN, Chen L, Thomas DJ, Timmerman LA, Nolan GP, Admon A, Crabtree GR (1994) Nature 369:497 b) Chuvpilo S, Avots A, Berberich-Siebelt F, Glockner J, Fischer C, Kerstan A, Escher C, Inashkina I, Hlubek F, Jankevics E, Brabletz T, Serfling E (1999) J Immunol 162:7294 c) Chuvpilo S, Zimmer M, Kerstan A, Glockner J, Avots A, Escher C, Fischer C, Inashkina I, Jankevics E, Berberich-Siebelt F, Schmitt E, Serfling E (1999) Immunity 10:261
54. a) Loh C, Shaw KT, Carew JA, Viola JP, Luo C, Perrino BA, Rao A (1996) J Biol Chem 271:10884 b) Loh C, Carew JA, Kim J, Hogan PG, Rao A (1996) Mol Cell Biol 16:3945
55. a) Timmerman LA, Clipstone NA, Ho SN, Northrop JP, Crabtree GR (1996) Nature 383:837 b) Shibasaki F, McKeon F (1995) J Cell Biol 131:735
56. Sugimoto T, Stewart S, Han M, Guan K-L (1998) EMBO J 17:1717
57. a) Matheos DP, Kingsbury TJ, Salma Ahsan U, Cunningham KW (1998) Genes Dev 11:3445 b) Stathopoulos AM, Cyert MS (1998) Genes Dev 11:3432 c) Stathopoulos-Gerontides A, Guo JJ, Cyert MS (1999) Genes Dev 13:798
58. Kayyali US, Zhang W, Yee AG, Seidman JG, Potter H (1997) J Neurochem 68:1668
59. For reviews see: a) Yakel JL (1997) Trends Pharmacol Sci 18:124 b) Snyder SH, Lai MM, Burnett PE (1998) Neuron 21:283 c) Genazzani AA, Carafoli E, Guerini D (1999) Proc Natl Acad Sci USA 96:5787
60. King MM, Huang CY, Chock PB, Nairn AC, Hemmings HC Jr, Chan KF, Greengard P (1984) J Biol Chem 259:8080
61 Ingebritsen TS, Cohen P (1983) Science 221:331
62. Fienberg AA, Hiroi N, Mermelstein PG, Song W, Snyder GL, Nishi A, Cheramy A, O'Callaghan JP, Miller DB, Cole DG, Corbett R, Haile CN, Cooper DC, Onn SP, Grace AA, Ouimet CC, White FJ, Hyman SE, Surmeier DJ, Girault J, Nestler EJ, Greengard P (1998) Science 282:838
63. a) Winder DG, Mansuy IM, Osman M, Moallem TM, Kandel ER (1998) Cell 92:25 b) Mansuy IM, Mayford M, Jacob B, Kandel ER, Bach ME (1998) Cell 92:39
64. Bito H, Deisseroth K, Tsien RW (1997) Curr Opin Neurobiol 7:419
65. Bito H, Deisseroth K, Tsien RW (1996) Cell 87:1203
66. a) Sussman MA, Lim HW, Gude N, Taigen T, Olson EN, Robbins J, Colbert MC, Gualberto A, Wieczorek DF, Molkentin JD (1998) Science 281:1690 b) Molkentin JD, Lu J-R, Antos CL, Markham B, Richardson J, Robbins J, Grant SR, Olson EN (1998) Cell 93:21
67. Chin ER, Olson EN, Richardson JA, Yang Q, Humphries C, Shelton JM, Wu H, Zhu W, Bassel-Duby R, Williams RS (1998) Genes Dev 12:2499
68. Dawson TM, Steiner JP, Dawson VL, Dinerman JL, Uhl GR, Snyder SH (1993) Proc Natl Acad Sci USA 90:9808
69. a) Sharkey J, Butcher SP (1994) Nature 371:46 b) Snyder SH, Sabatini DM, Lai MM, Steiner JP, Hamilton GS, Suzdak PD (1998) Trends Pharmacol Sci 19:21
70. Lee JP, Palfrey HC, Bindokas VP, Ghadge GD, Miller RJ, Roos RP (1999) Proc Natl Acad Sci USA 96:3251
71. Terashima A, Nakai M, Kawamata T, Taniguchi T, Yasuda M, Maeda K, Tanaka C (1998) Brain Res 786:255
72. Bennett PC, Singaretnam LG, Zhao WQ, Lawen A, Ng KT (1998) FEBS Lett 431:386
73. Peterson LB, Cryan JG, Rosa R, Martin MM, Wilusz MB, Sinclair PJ, Wong F, Parsons JN, O'Keefe SJ, Parsons WWH, Wyvratt M, Sigal NH, Williamson AR, Wiederrecht GJ (1998) Transplantation 65:10
74. Renoir JM, Mercier-Bodard C, Hoffman K, Le Bihan S, Ning Y-M, Sanchez ER, Handschumacher RE, Baulieu EE (1995) Proc Natl Acad Sci USA 92:4977
75. Huse M, Chen YG, Massague J, Kuriyan J (1999) Cell 96:425
76. Chambraud B, Radanyi C, Camonis JH, Rajkowski K, Scumacher M, Baulieu EE (1999) Proc Natl Acad Sci USA 96:2104
77. Aramburu J, Yaffe MB, Lopez-Rodriguez C, Cantley LC, Hogan PG, Rao A (1999) Science 285:2129
78. Spitzfaden C (1993) Diss. ETH (No10406) Zurich
79. Merritt EA, Bacon DJ (1997) Methods Enzymol 277:505

The Ca^{2+} Pump of Sarcoplasmic and Endoplasmic Reticulum Membranes

Giuseppe Inesi[1,*], Chikashi Toyoshima[2]

[1] Department of Biochemistry and Molecular Biology, University of Maryland School of Medicine, Baltimore, MD 21201, USA (*E-mail: ginesi@umaryland.edu*)
[2] Institute of Molecular and Cellular Biosciences, The University of Tokyo, Bunkyo-ku, Tokyo 113, Japan (*E-mail: ct@iam.u-tokyo.ac.jp*)

The sarco-endoplasmic reticulum Ca^{2+} (SERCA) ATPases constitute a family of enzyme isoforms whose function is to sequester cytosolic Ca^{2+} by active transport across intracellular membranes. Coupling of ATP hydrolysis and Ca^{2+} transport is based on an intramolecular long-range linkage between ATP hydrolysis within the cytosolic region, and vectorial displacement of bound Ca^{2+} within the membrane-bound region of the ATPase. Equilibrium and kinetic measurements, mutational analysis and structural studies have produced a considerable body of information on this system. That information is discussed here in order to unveil relationships of structure and function within the ATPase protein, and to clarify the coupling mechanism of ATP hydrolysis and Ca^{2+} transport at the molecular level.

Keywords. SERCA ATPases, Ca^{2+} transport, Sarcoplasmic reticulum, Protein structure and function, Calcium ion

1 Introduction

The microsomal fraction of muscle homogenates was originally found to prevent contractile activation of myofibrils upon addition of ATP and, for this reason, was referred to as the "relaxing factor". The microsomes were identified as vesicular fragments of sarcoplasmic reticulum [1], and the relaxing activity was attributed to Ca^{2+} transport and sequestration by an ATPase associated with the membrane vesicles [2, 3]. While the Ca^{2+} ATPase is a major component of the sarcoplasmic reticulum of muscle cells, lower concentrations of the same enzyme or its isoforms are also present in the endoplasmic reticulum of other eukaryotic cells. For this reason the enzyme is referred to as *S*arco-*E*ndoplasmic *R*eticulum *Ca*$^{2+}$ ("SERCA") ATPase. The SERCA1a isoform and its SERCA1b splicing variant are expressed in adult and fetal fast twitch skeletal muscle [4–6]. The SERCA2a isoform is prevalent in cardiac and slow twitch skeletal muscle, and its SERCA2b splicing variant in multiple tissues [7–9]. Furthermore, two SERCA3 isoforms have been identified in endothelial, epithelial and hematopoietic cells [10–12]. The SERCA ATPase plays a very important role in cell physiology inasmuch as it allows sequestration of cytosolic Ca^{2+} that can be subsequently released to trigger various signaling mechanisms [13].

SERCA1a ATPase accounts for a large percentage of the total protein content of sarcoplasmic reticulum vesicles isolated from fast skeletal muscle (Fig. 1). This turns out to be of great experimental advantage in studies of function and structure. When negatively stained vesicles are examined by electron microscopy, the ATPase molecules appear as densely spaced granules on the outer surface of the membrane, therefore facing the cytosol in the intact cell (Fig. 1). The outer granules are connected through thin stalks to membrane-bound regions of the ATPase. The membrane-bound regions appear as particles on the outer membrane leaflet of freeze-fractured vesicles (Fig. 1). Such a distribution of protein on the cytosolic side and membrane bilayer is consistent with electron density profiles obtained by low-angle X-ray diffraction studies of stacked vesicles [14].

2 Functional Characterization

The overall function of SERCA ATPase is to couple hydrolysis of ATP to active transport of Ca^{2+}, resulting in accumulation of medium Ca^{2+} in the lumen of the vesicles. The ATPase is activated by micromolar medium Ca^{2+}, and is inhibited by mM lumenal Ca^{2+}. Therefore, the pump is kinetically competent within this concentration range and, under favorable conditions, is able to produce a three- to four-order of magnitude concentration gradient between sub-micromolar medium Ca^{2+} and mM lumenal Ca^{2+}. Since two Ca^{2+} are transported per ATPase cycle, the free energy requirement against such a gradient is:

$$\begin{aligned}\Delta G &= 2RT\ln(Ca^{2+}_{in}/Ca^{2+}_{out})\\ &= 2RT\ln 10^{3\text{ or }4} \cong 7 \text{ or } 10 \text{ kcal per cycle per mole of ATPase}\end{aligned}$$

This much free energy can be derived from hydrolysis of one ATP ($\Delta G = 7.3$ kcal/mol under standard conditions) in media of physiological composition.

The coupling mechanism of ATP hydrolysis and Ca^{2+} transport was previously subjected to detailed analysis in terms of sequential steps and their microscopic

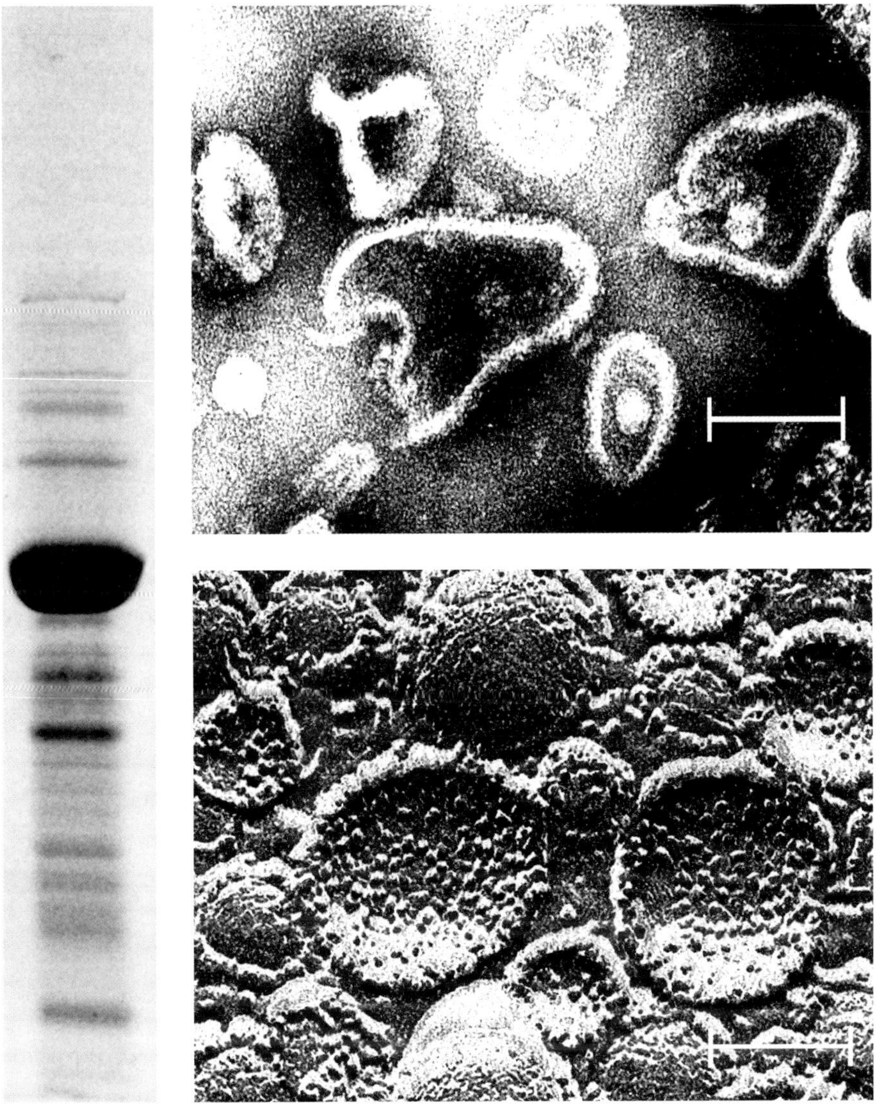

Fig. 1. High SERCA ATPase content of native SR vesicles. The high ATPase content is shown by the heavy band on the electrophoretic gel of the microsomal fraction on the left side of the figure. The ATPase molecules appear as densely packed particles on the outer surface of negatively stained vesicles (*right side, top*). They also appear on the outer leaflet (inner surface) of freeze-fractured vesicles as particles intruding the membrane bilayer (*right side, bottom*). Magnification mark = 0.1 μm. Figure derived from Inesi and Scales [53]

constants [15]. Here, we describe a minimal number of sequential reactions that shall be considered again later regarding their topology within the protein structure. The reactions to be considered are the following:

1. *Ca^{2+} binding.* ATPase activation requires initial binding of medium Ca^{2+} to the enzyme. Binding occurs in the absence of ATP, and can be measured under equilibrium conditions. The resulting isotherm demonstrates high affinity and cooperative binding of two Ca^{2+} per ATPase. The Ca^{2+} saturation curve shown in Fig. 2 is a fundamental parameter, comparable to the oxygen saturation curve of hemoglobin, but with an additional linkage to catalytic activation. In fact, the Ca^{2+} concentration (0.2–2.0 μM) dependence of binding matches closely the Ca^{2+} concentration dependence of enzyme and transport activation. The state of the enzyme activated by Ca^{2+} has been widely referred to in the literature as "E1".
2. *Formation of phosphorylated enzyme intermediate.* When ATP is added to Ca^{2+}-activated ATPase, the ATP terminal phosphate is rapidly transferred to the enzyme, forming a covalent phosphorylated intermediate [16]. In parallel with enzyme phosphorylation, two bound Ca^{2+} per phosphoenzyme become unavailable for isotopic exchange on the cytosolic surface of the membrane (Fig. 3). The stoichiometric and kinetic correspondence of enzyme phosphorylation and Ca^{2+} occlusion is consistent with a causal relationship of the two phenomena, and with their identity as intermediate steps of the catalytic and transport mechanism.
3. *Vectorial dissociation of bound Ca^{2+}.* It is possible to demonstrate by rapid kinetic experimentation and calcium isotopic chase that the two Ca^{2+} bound to the enzyme are released into the lumen of the vesicles before hydrolytic cleavage of the phosphorylated enzyme intermediate [17]. It is then apparent that, as a consequence of phosphorylation, the Ca^{2+}-binding sites undergo a reduction of binding strength and a vectorial change in orientation. The state of the enzyme exhibiting low affinity and lumenal orientation of the Ca^{2+} sites is referred to in the literature as "E2". The Ca^{2+} concentration dependence of ATPase inhibition and phosphoenzyme reversal to ATP suggests that, in the "E2" state, the Ca^{2+}-binding sites have a dissociation constant in the mM range [18] as opposed to the micromolar range in the "E1" state.

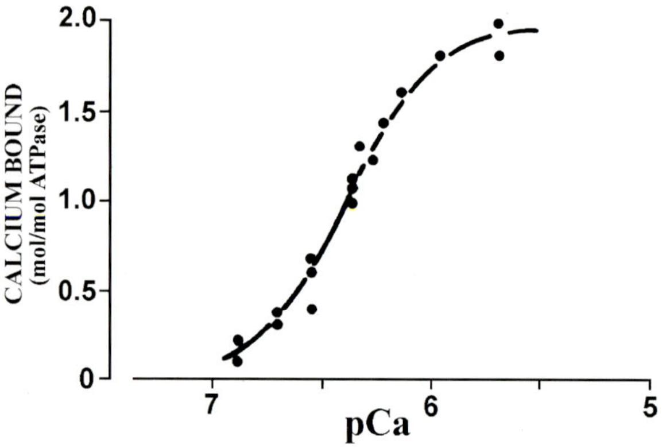

Fig. 2. Equilibrium isotherm of Ca^{2+} binding to the SERCA ATPase of native SR vesicles. Binding was measured by equilibration of radioactive isotope in size exclusion chromatography columns, in the absence of ATP, and in the presence of mM Mg^{2+} and Ca-EGTA buffers to yield various free Ca^{2+} concentrations. The experimental points were fitted assuming cooperative binding of two Ca^{2+}. A single Ca^{2+}-binding equation did not yield a satisfactory fit. Figure derived from Inesi et al. [54]

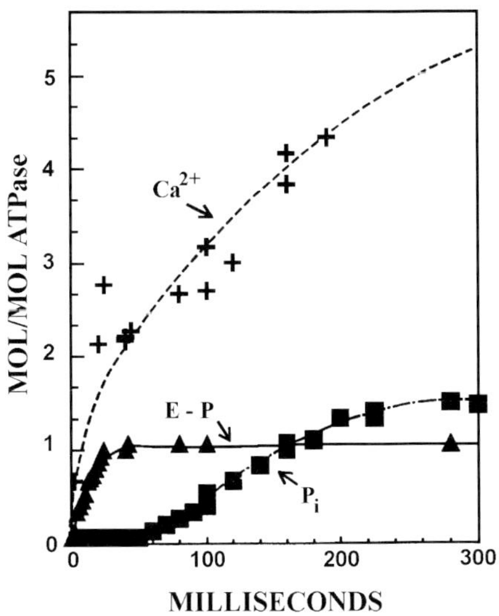

Fig. 3. Sequence of pre-steady state reactions following addition of ATP to SR vesicles preincubated with Ca^{2+}. Note the rapid formation of stoichiometric levels of phosphorylated enzyme intermediate, and the simultaneous occlusion of two moles of calcium per mole of phosphorylated enzyme. Hydrolytic cleavage of Pi occurs following a lag period, and then steady state cycling begins, maintaining a 2:1 molar ratio between Ca^{2+} transport and Pi production. The measurements were obtained at 25 °C with the aid of a quench-flow apparatus. Figure adapted from Inesi et al. [55]

4. *Hydrolytic cleavage of the phosphoenzyme.* Production of Pi by hydrolytic cleavage of the phosphoenzyme occurs following a rate limiting step corresponding to vectorial displacement of bound Ca^{2+}, as can be clearly seen in presteady state experiments (Fig. 3). The hydrolytic reaction is readily reversible, as the enzyme can be phosphorylated again by Pi which is then transferred to ADP if present, thereby yielding Pi/ATP exchange [19]. An experimentally interesting aspect of this reversal is the enzyme reaction with Pi in the absence of Ca^{2+} and nucleotide [20]. The Pi reaction was studied under equilibrium conditions (i.e., in the absence of enzyme turnover), and found to be strongly inhibited by Ca^{2+} binding to the high affinity sites of the ATPase. These experiments demonstrate that the ATPase can bind either Ca^{2+} or Pi, but not both at the same time. In the presence of Ca^{2+}, enzyme phosphorylation requires the free energy of ATP. The consequent phosphoryl transfer to the enzyme causes dissociation of bound Ca^{2+}. Such is the fundamental feature of energy transduction in this enzyme.

3
Structural Characterization

The SERCA ATPase is an approximately 100 kD protein that appears to be catalytically active in its monomeric state. The amino acid sequence of the SERCA1a isoform was obtained in part by direct analysis [21], and then in toto by cDNA cloning and sequencing [6]. The amino acid sequences of other SERCA isoforms were subsequently derived from their cDNA clones, revealing extensive homology and only slight changes in size. The original two-dimensional model for the sequence distribution with respect to the membrane bilayer includes short amino and carboxyl ends protruding on the cytosolic side, a smaller and a larger cytosolic loop, five stalk segments, and ten transmembrane segments with very small loops on the lumenal side (Fig. 4). Reaction of antibodies with specific epitopes [22, 23] is in agreement with the topology proposed by the model.

Images of the SERCA ATPase three-dimensional structure were derived from electron microscopy of ordered arrays. Consistent with previous evidence obtained by direct microscopic visualization, the three-dimensional images reconstructed from ordered arrays show a globular region outside the cytosolic surface of the membrane, connected through a stalk segment to a membrane spanning region [24]. Higher resolution and much improved images were then obtained by cryoelectron microscopy [25]. Two types of preparations have been examined [26, 27]: One type derives directly from sarcoplasmic reticulum vesicles yielding tubular crystals of ATPase molecules stabilized by decavanadate in the absence of Ca^{2+}. The other type, formed in the presence of Ca^{2+}, consists of multilamellar crystals of ATPase molecules purified following solubilization of SR vesicles with detergent.

The two preparations yield different images (Fig. 5). The differences are more obvious in the cytosolic region, but may also involve the transmembrane region [27]. The image derived from tubular crystals shows a single cytosolic domain, while the image derived from multilamellar crystals shows two distinct cytosolic domains of different mass. The difference between the images obtained from the two preparations was attributed to the absence or the presence of Ca^{2+}, therefore corresponding to an inactive or active state, respectively. An alternative explanation is that the image obtained from the preparation stabilized with decavanadate is influenced by

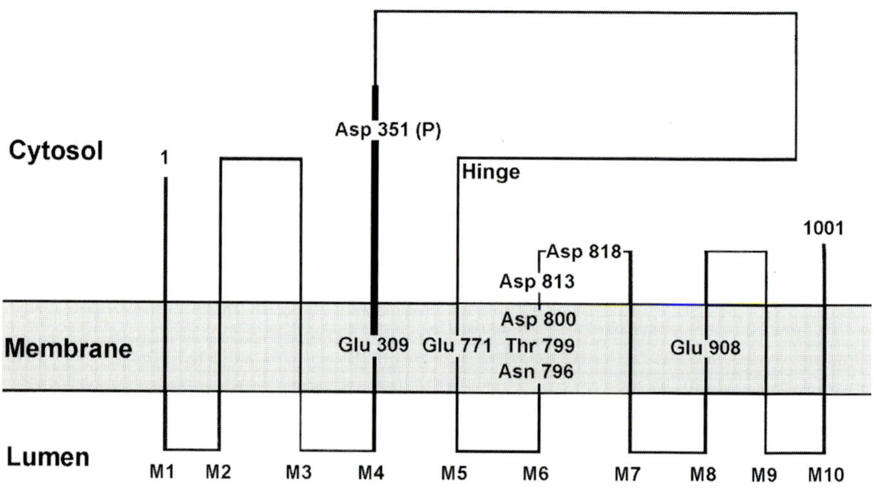

Fig. 4. Two-dimensional diagram of the SERCA1a ATPase amino acid sequence distribution with respect to the membrane bilayer. The residue undergoing phosphorylation (asp351) is shown in the large cytosolic loop. The six residues believed to participate directly in Ca^{2+} binding are shown in the M4, M5, M6 and M8 transmembrane segments [37]. The two residues shown on the cytosolic extension of M6 appear to influence Ca^{2+} binding [56]. The thick line shows a segment of prominent homology in several cation transport ATPases. The diagram is derived from the original model of MacLennan et al. [6]

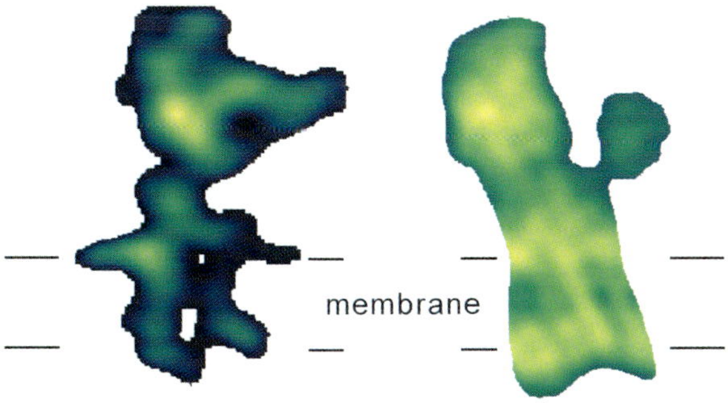

Fig. 5. Images of SERCA ATPase derived by cryoelectron microscopy of tubular (*left*) or multilamellar (*right*) crystals. The tubular crystals were obtained in the absence of Ca^{2+} and in the presence of decavanadate, from native SR vesicles. The multilamellar crystals were obtained in the presence of Ca^{2+} and in the absence of vanadate, from enzyme solubilized with detergent and purified in the presence of Ca^{2+} [27]

formation of an ATPase complex with orthovanadate, corresponding to a transition state analog of the phosphorylation reaction. Whether Ca^{2+} induced or vanadate induced, these studies demonstrate that the ATPase can acquire different conformations corresponding to distinct functional states.

4
Relationships of Structure and Function

Identification of functional domains and amino acid side chains within the three-dimensional structure of the enzyme is required to understand the coupling mechanism of ATP hydrolysis and cation transport.

4.1
Nucleotide Binding and Phosphorylation Domain

Modeling based on limited analogies with adenylate kinase [28] and more extensive analogies with halogenases [29] indicate that the segments of the large cytosolic loop (Fig. 5) are folded in the three-dimensional structure of the enzyme to accommodate nucleotide binding. Experimentally, it was found that residues located wide apart on the amino acid sequence (Lys492 to Lys684) contribute to ATP binding [30, 31]. Furthermore, derivatization of Lys515 with FITC [32] interferes with ATP binding although still permitting binding and utilization of other substrates such as acetyl phosphate and p-nitrophenyl phosphate. This is consistent with the proximity of Lys515 to the adenine-binding moiety of the ATP site. Location of the nucleotide-binding site in a groove of the cytosolic region was demonstrated [33] by electron microscopy of tubular crystals of ATPase with bound CrATP (Fig. 6).

The ATPase residue undergoing phosphorylation as an intermediate step of ATP utilization is Asp351 [34, 35]. This residue resides in the large cytosolic loop of the ATPase sequence (Fig. 4), and is flanked by residues often found in catalytic

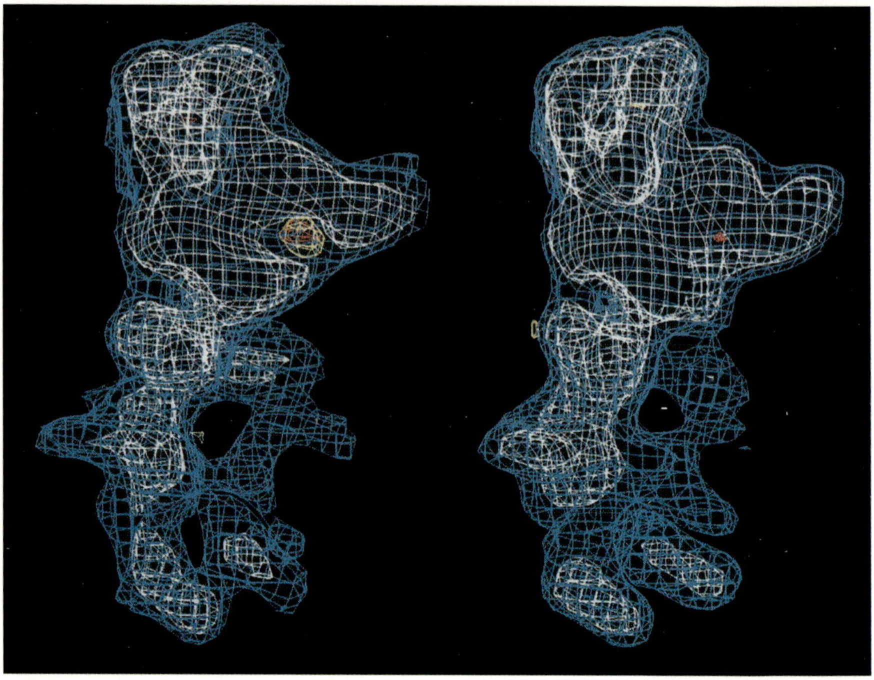

Fig. 6. Nucleotide-binding site in the cytosolic region of SERCA ATPase. Tubular crystals of SERCA ATPase were stabilized with decavanadate, using enzyme pre-incubated in *a* absence or *b* in the presence of a non-hydrolyzable analog of ATP (CrATP). The two preparations were then examined by cryoelectron microscopy, and the images were compared. In these fishnet representations the blue and white lines enclose 75 and 35% of the expected volume, respectively. The yellow and red nets indicate the location of density differences attributed to the presence of bound CrATP. The red net corresponds to a 99.5% statistical confidence in *a*, and 99.9% in *b* [33]

phosphorylation sites [29]. In the folded ATPase structure, Cys364, Cys498 and Cys636 are apparently close to the phosphorylation site since they are protected by orthovanadate from derivatization with -SH reagents [36].

4.2 Ca^{2+}-Binding Domain

It was originally noted by Clarke et al. [37] that single mutations of any of six specific amino acids in the M4, M5, M6 and M8 transmembrane segments (Fig. 4) produce ATPase inhibition. An interesting aspect of these mutants is that the enzyme remains capable of being phosphorylated by Pi, but the Pi reaction is not inhibited by Ca^{2+} as normally observed with wild-type enzyme. On the basis of these findings it was suggested that these mutations interfere with Ca^{2+} binding [37–40], whereby activation of ATP utilization and inhibition of the Pi reaction would not occur due to lack of bound Ca^{2+}. The original suggestion of Ca^{2+}-binding defects was based on indirect evidence, since direct measurements of Ca^{2+} binding were prevented by limiting quantities of mutated protein. The recent development of recombinant adenovirus with SERCA cDNA as an efficient vector for transfer into cultured cells [41] has greatly increased the yield of recombinant protein, thereby permitting direct measurements of Ca^{2+} binding by wild-type enzyme and mutants of interest. It has since been possible to demonstrate [42] that, in the presence of μM Ca^{2+}, and in the absence of ATP, mutations of Glu309 (M4), Asp771 (M5), Asn796 (M6), Thr799 (M6), Glu800 (M6) and Glu908 (M8) inhibit partially or totally the specific Ca^{2+} binding which is involved in enzyme activation. Kinetic and spectroscopic [43, 44] studies indicate that the carboxylic groups of these residues undergo changes in pK and/or hydrogen bonding whereby Ca^{2+}/H^+ exchange and counter-transport occur in parallel with ATP utilization.

Molecular modeling of M4, M5, M6 and M8 transmembrane clusters indicate that the oxygen functions of the six amino acids in question can be oriented to yield discrete areas of electronegativity that are consistent with binding of two calcium ions [40]. The actual relationship of the M4, M5, M6 and M8 in the transmembrane cluster, shown in Fig. 7, has been established by high-resolution structural analysis of three-dimensional crystals obtained in the presence of Ca^{2+} [45]. While the four segments are predominantly helices, the middle portions of M4 and M6 are likely to include flexible segments which would facilitate participation of all three M6 residues (Asn796, Thr799 and Asp800) believed to be involved in Ca^{2+} binding [45, 46].

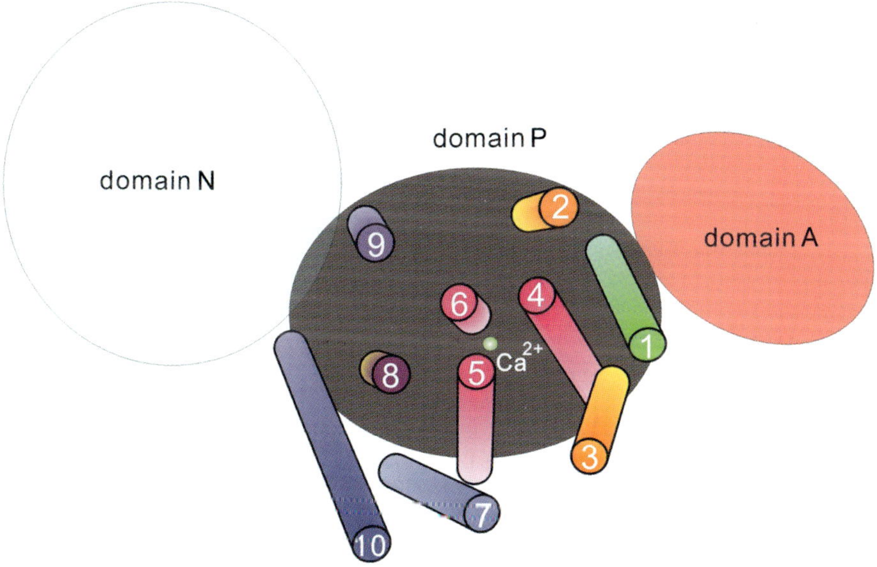

Fig. 7. Relationships of cytosolic and transmembrane domains of SERCA ATPase viewed from the cytosolic side of the membrane. This diagram is derived from on-going diffraction studies of three-dimensional crystals [45]. Note the close relationship of transmembrane segments M4, M5 and M6, and the neighboring location of M8. Note also that the diffraction studies identify three distinct domains (A, P, N) in the cytosolic region. Domain P is the phosphorylation domain

5
Long-Range Intramolecular Linkage of Functional Domains

Occupancy of Ca^{2+} sites is very important to the stability of the ATPase structure, since, in the absence of Ca^{2+}, the solubilized enzyme is highly susceptible to denaturation. This indicates that cooperative interaction of Ca^{2+} with oxygen functions of the transmembrane segments confers stability to the entire protein. From the functional standpoint, occupancy of the Ca^{2+} sites is required for enzyme activation, since, in the absence of Ca^{2+}, the catalytic site cannot utilize ATP. Conversely, and concomitant with each enzyme cycle, Ca^{2+} binding is transiently destabilized by phosphorylation of the catalytic site upon utilization of ATP. Considering a distance of 40–50 Å between Ca^{2+} and catalytic sites, it is apparent that a long-range intramolecular linkage is involved in determining structural and functional states in this enzyme [47].

Sequence alignment of several enzymes that couple cation transport and hydrolysis of ATP through formation of a phosphorylated intermediate ("P-ATPases") reveals prominent homologies in the stalk and membrane segments (S4 and M4) connecting the phosphorylation site with the Ca^{2+} (or other cations) binding domain. Such homologies suggest that these segments may be a common device in the long-range linkage of functional domains. In fact, detailed mutational analysis of the SERCA ATPase shows that several mutations produce inhibition of catalytic turnover, as indicated by the slower decay of phosphorylated enzyme intermediate obtained with ATP in the presence of Ca^{2+}, or with Pi in the absence of Ca^{2+} [39, 40]. Such mutations may be in selected areas of the cytosolic region of the ATPase, but more often in the stalk and transmembrane segments. It is then apparent that stalk and transmembrane segments (most prominently M4, M5 and M6) are clustered in a bundle stabilized by multiple cooperative interactions, whereby single amino acid mutations may produce structural readjustments that result not only in interference with Ca^{2+} binding but also with catalytic functions.

A further indication of long-range linkage is provided by the topology of ATPase sensitivity to thapsigargin, a sesquiterpene lactone [48] derived from the plant *Thapsia garganica*, and a highly specific inhibitor of SERCA ATPases. Stoichiometric interaction of thapsigargin with the enzyme yields a "global" effect that favors stabilization of crystalline protein arrays [49, 50], and inhibits ATP or Pi utilization as well as Ca^{2+} binding [51]. Mutational analysis shows that the site of thapsigargin interaction with the ATPase is within or near the S3 stalk segment and the membrane interface [52]. Therefore, the global effect of thapsigargin is produced through perturbation of stalk segments that are involved in the long-range linkage of catalytic and Ca^{2+}-binding domains.

6
Conclusions

The natural abundance of native protein, the availability of recombinant methods for mutational analysis, and the increasing resolution of structural images, have all contributed to a detailed characterization of structure and function in the SERCA

ATPase. Thermodynamic and kinetic characteristics of the catalytic and transport cycle have been defined. The topology of functional domains and their linkage within the protein structure have been characterized. It is expected that the ongoing X-ray crystallographic study will unveil details of the structure at the atomic level.

References

1. Ebashi S, Lippman F (1962) J Cell Biol 14:389
2. Hasselbach W, Makinose M (1961) Biochem Z 333:518
3. Hasselbach W, Makinose M (1963) Biochem Z 339:94
4. Brandl CJ, Green NM, Korczak B, MacLennan DH (1986) Cell 44:597
5. Brandl CJ, DeLeon S, Martin DR, MacLennan DH (1987) J Biol Chem 262:3768
6. MacLennan DH, Brandl CJ, Korczak B, Green NM (1985) Nature 316:696
7. Arai M, Otsu K, MacLennan DH, Periasamy M (1992) Am J Physiol Cell Physiol 262:C614
8. Gunteski-Hamblin AM, Greeb J, Shull GE (1988) J Biol Chem 263:15032
9. Lytton J, Zarain-Herzberg A, Periasamy M, MacLennan DH (1989) J Biol Chem 264:7059
10. Anger M, Samuel JL, Marotte F, Wuytack F, Rappaport L, Lompre AM (1994) J Mol Cell Cardiol 26:539
11. Burk SE, Lytton J, MacLennan DH, Shull GE (1989) J Biol Chem 264:18561
12. Wu KD, Lee WS, Wey J, Bungard D, Lytton J (1995) Am J Physiol Cell Physiol 269:C775
13. Hussain A, Inesi G (1999) J Memb Biology, in press
14. Herbette LG, Marquardt J, Scarpa A, Blasie JK (1977) Biophys J 20:245
15. Inesi G, Kurzmack M, Lewis D (1988) Methods Enzymol 157:154
16. Yamamoto T, Tonomura Y (1967) J Biochem (Tokyo) 62:558
17. Inesi G (1987) J Biol Chem 262:16338
18. deMeis L, Carvalho MGC (1971) Biochemistry 13:5032
19. Makinose M, Hasselbach W (1971) FEBS Lett 121:271
20. Masuda H, deMeis L (1973) Biochemistry 12:4581
21. Allen G, Trinnaman BJ, Green NM (1980) Biochem J 187:591
22. Matthews I, Colyer J, Mata AM, Green NM, Sharma RP, Lee AG, East JM (1989) Biochem Biophys Res Commun 161:683
23. Clarke DM, Loo TW, MacLennan DH (1990) J Biol Chem 265:17405
24. Taylor KA, Dux L, Martonosi A (1986) J Mol Biol 187:417
25. Toyoshima C, Sasabe H, Stokes DL (1993) Nature 362:469
26. Zhang P, Toyoshima C, Yonekura K, Green NM, Stokes DL (1998) Nature 392:835
27. Ogawa H, Stokes DL, Sasabe H, Toyoshima C (1998) Biophys J 75:41
28. Green NM (1989) Biochem Soc Trans 17:972
29. Aravind L, Galperin MY, Koonin EV (1998) TIBS 23:127
30. Yamamoto H, Imamura Y, Tagaya M, Fukui T, Kawakita M (1989) J Biochem (Tokyo) 106:1121
31. McIntosh DB, Woolley DG, Berman MC (1992) J Biol Chem 267:5301
32. Mitchinson C, Wilderspin A, Trinnaman BJ, Green NM (1982) FEBS Lett 146:87
33. Yonekura K, Stokes DL, Sasabe H, Toyoshima C (1997) Biophys J 72:997
34. Bastide F, Meissner G, Fleischer S, Post RL (1973) J Biol Chem 248:8385
35. Degani C, Boyer PD (1973) J Biol Chem 248:8222
36. Hua S, Fabris D, Inesi G (1999) Biophys J 77:2217
37. Clarke DM, Loo TW, Inesi G, MacLennan DH (1989) Nature 339:476
38. Andersen JP, Vilsen B (1992) J Biol Chem 267:19383
39. Andersen JP (1995) Biosci Rep 15:243
40. Chen L, Sumbilla C, Lewis D, Zhong LL, Strock C, Kirtley ME, Inesi G (1996) J Biol Chem 271:10745
41. Inesi G, Lewis D, Sumbilla C, Nandi A, Strock C, Huff KW, Rogers TB, Johns DC, Kessler PD, Ordahl CP (1998) Am J Physiol 274:C645
42. Strock C, Cavagna M, Peiffer W, Sumbilla C, Lewis D, Inesi G (1998) J Biol Chem 273:15104
43. Yu X, Hao L, Inesi G (1994) J Biol Chem 269:16656

44. Barth A, Kreutz W, Mäntele W (1994) Biochim Biophys Acta Bio Membr 1194:75
45. Toyoshima C, Nakasako M, Nomura H, Ogawa H (2000) Nature, in press
46. Soulié S, Neumann JM, Berthomieu C, Moller JV, Le Maire M, Forge V (1999) Biochemstry 38:5813
47. Inesi G, Cantilina T, Yu X, Nikic D, Sagara Y, Kirtley ME (1992) Ann NY Acad Sci 671:32
48. Christensen SB (1988) Acta Chem Scand B42:623
49. Sagara Y, Wade JB, Inesi G (1992) J Biol Chem 267:1286
50. Stokes DL, Lacapère JJ (1994) J Biol Chem 269:11606
51. Sagara Y, Fernandez-Belda F, deMeis L, Inesi G (1992) J Biol Chem 267:12606
52. Zhong L, Inesi G (1998) J Biol Chem 273:12994
53. Inesi G, Scales DJ (1974) Biochemistry 13:3299
54. Inesi G, Kurzmack M, Coan C, Lewis D (1980) J Biol Chem 255:3025
55. Inesi G, Watanabe T, Coan C, Murphy AJ (1982) Ann NY Acad Sci 402:515
56. Menguy T, Corre F, Bouneau L, Deschamps S, Moller JV, Champeil P, le Maire M, Falson P (1998) J Biol Chem 273:20134

The Plasma Membrane Ca^{2+} ATPase

Danilo Guerini

Laboratory of Biochemistry, Swiss Federal Institute of Technology (ETH), 8092 Zürich, Switzerland (*E-mail: danilo.guerini@pharma.novartis.com*)
Present address: Department of Metabolic and Cardiovascular Diseases, Novartis Pharma AG, WKL125, 10.57, Klybeckstrasse 141, Basel 4057, Switzerland

The plasma membrane Ca^{2+} ATPase (PMCA) or Ca^{2+} pump transports Ca^{2+} ions out of the cells by using the energy stored in ATP. The plasma membrane Ca^{2+} pump has been found not only in all mammalian cells, but also in varieties of invertebrates and in yeast. So far no PMCA genes have been found in bacteria. The biochemical and biophysical properties of the pump have been primarily studied on the enzyme purified from the membrane of the red blood cells but, recently, molecular biological methods have been successfully used to study the properties of the different PMCA isoforms. The plasma membrane pump is encoded by four genes and alternative splicing at two independent sites increases the number of the possible isoforms: transcripts for more than 20 isoforms have been detected. Some of the PMCA variants have a different affinity to calmodulin. The activity of these alternatively spliced pumps is possibly differently regulated by kinase-mediated phosphorylation. Here an update will be presented on the regulation of the expression of PMCA genes and splice variants, on the subcellular distribution of the protein and on the mechanism by which the PMCA pump translocates Ca^{2+} through the plasma membrane. Genetic evidence indicates that PMCA2, one of the isoforms expressed at high level only in brain, is responsible for hearing defects in mice.

Keywords. Ca^{2+} transport, Calmodulin and Ca^{2+}-dependent ATPase, Ca^{2+} pump, Gene regulation, Calcineurin

List of Abbreviations

PMCA	plasma membrane Ca^{2+} ATPase
SERCA	sarcoplasmic reticulum/endoplasmic reticulum Ca^{2+} ATPase
PM	plasma membrane
ER	endoplasmic reticulum
TM	transmembrane
FITC	fluorescein isothiocyanate
dfw	deaf-waddler

1 Introduction: General Properties of the Pump

The maintenance of a low free Ca^{2+} concentration is vital to the correct functioning and survival of the cell [1]. The opening of specific channels in the plasma membrane allows Ca^{2+} to enter cells down a large concentration gradient: in the cytosol free Ca^{2+} oscillates at around values of 0.2 μM, in the extracellular space around 1–2 mM. Efficient systems have evolved to transport the Ca^{2+} out of the cytosol across the plasma membrane and into subcellular organelles [1]. Three proteins perform this function: the Na^+/Ca^{2+} exchanger and the Ca^{2+} ATPases (pumps) of the plasma membrane and of the sarcoplasmic/endoplasmic reticulum. While the exchanger couples the transport of Ca^{2+} to the "downhill" cotransport of Na^+ [2], Ca^{2+} pumps use the energy stored in ATP to extrude the Ca^{2+} from the cell [1, 3].

The plasma membrane calcium ATPase has an essential function in non-muscle cells, whereas its importance in muscle cells is likely to be minor compared to that of the much more abundant SERCA pump. Nevertheless, the small amounts of Ca^{2+} re-exported by the PMCA Ca^{2+} pump in muscle cells are physiologically relevant, since the manipulation of the concentration of the PMCA pump by genetic engineering influences the resting level of cytosolic Ca^{2+} [4].

The plasma membrane Ca^{2+} pump belongs to the ion motive ATPase (pump) family [3, 5]. Their common property is the formation of an energized enzyme intermediate from ATP which is required for the translocation of the ion: the γ-phosphate of ATP is transferred to an Asp in the active site resulting in the formation of a high-energy acyl phosphate. The reaction is coupled to the transition of the enzyme between two conformational states termed E_1 and E_2 (Fig. 1) [6].

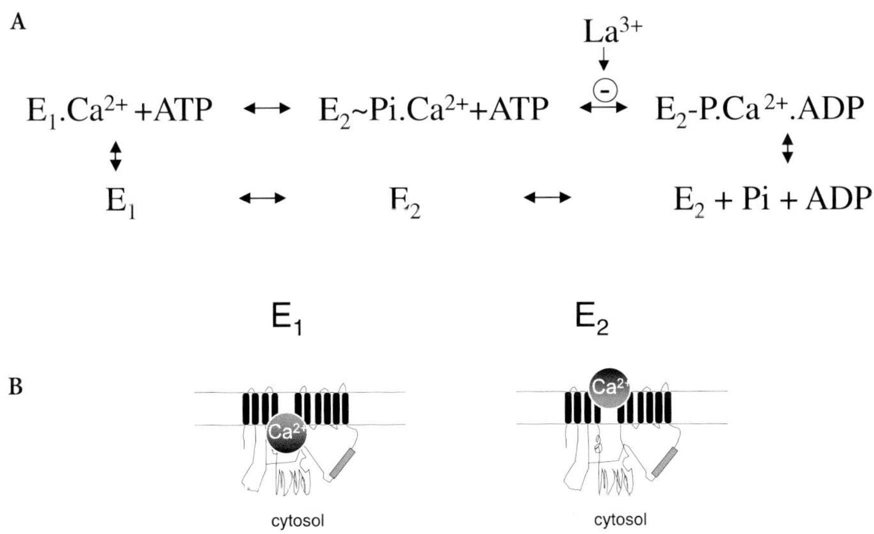

Fig. 1. Catalytic cycle of the PMCA pump. **A** The signs "~" and "-" indicate the high- and low-energy content of the complex between the enzyme and the phosphate atom. The two major conformational states of the pump (E_1 and E_2) are indicated. The site where La^{3+} is predicted to inhibit the decay of the phosphoenzyme intermediate (and therefore increase the amount of the intermediate) is indicated by (−). **B** Simplistically, in the E_1 state the high affinity Ca^{2+}-binding site is open toward the cytosol and in the E_2 state toward the extracellular space

As in all other P-type ATPases, the PMCA pump is inhibited by μM concentrations of the phosphate analogue orthovanadate $[VO_3(OH)]^{2-}$. The other general inhibitor of P-type ATPases, La^{3+}, inhibits the Ca^{2+} transport by the PMCA but, in contrast to all other P-type pumps, it stabilizes its phosphorylated intermediate by inhibiting its decay [7]. This property can be used to identify the phosphorylated intermediate of the PMCA pump in crude membrane preparations of tissues. The PMCA pump transports Ca^{2+} with a stoichiometry of 1 Ca^{2+} per 1 ATP, which differs from the Ca^{2+} pump of sarcoplasmic reticulum, whose optimal Ca^{2+}/ATP stoichiometry is 2 [8].

2 Structure of the Pump

2.1 Characterization of the PMCA Pump

An ATP-dependent Ca^{2+} transport system was described in erythrocytes more than 30 years ago [9]. The pump was purified as a protein of molecular mass about 134,000 Da from human erythrocytes after its ability to bind to calmodulin had been recognized [10]. The reconstitution in artificial liposomes of the purified pump enabled insights into its biochemical properties [11]. The calmodulin-binding domain was identified with the help of a photoactivatable derivative of calmodulin [12]. This domain, whose length is about 30 amino acids, has the propensity to form an amphiphilic helix of basic and apolar amino acids (Fig. 2). This property has subsequently been observed for the other calmodulin-binding domains of other calmodulin-modulated proteins.

2.2 Primary Structure of the Pump

The sequence of rat PMCA1 and 2 and human PMCA1 pumps was determined by cloning procedures in 1988 [13, 14]. A third rat (PMCA3) [15] and a fourth human (PMCA4) isoform [16] were characterized soon after. Further reports have

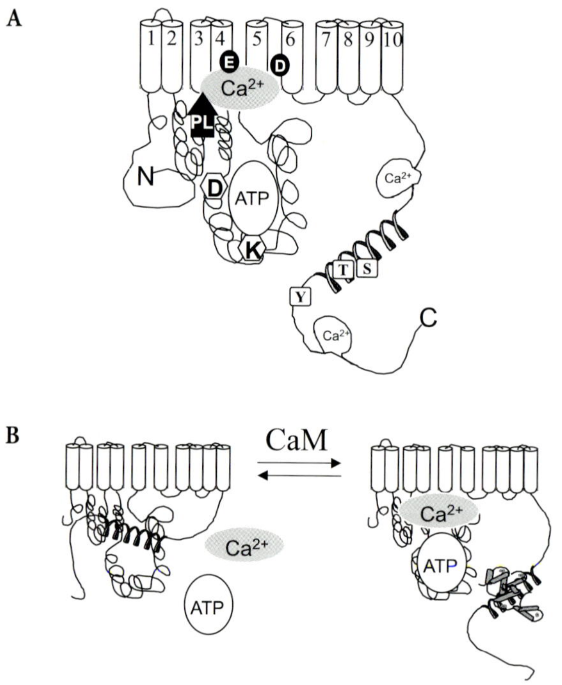

Fig. 2. Model of the membrane topology of the PMCA pump and the effect of calmodulin. (**A**) Model is based on the sequence of the PMCA4CI protein. D(=D465), the residue phosphorylated during the catalytic cycle, and K(=K591), the residue labelled by FITC in the ATP-binding domain, are surrounded by a hexagon. The location of the ATP-binding domain is shown (ATP). The Thr and the Ser residues located in the calmodulin-binding domain, phosphorylated by protein kinase C and A, respectively, and the Tyr, that was suggested to be phosphorylated by pp60(src), are shown in bold (*square*). One of the two acidic phospholipid-binding domains (PL, the other phospholipid-binding domain is the C-terminal calmodulin-binding site), the calmodulin-binding domain (helix) are also indicated. The amino acids in the transmembrane domains indicated by black circles are involved in the transport of Ca^{2+}: E[=Glu423 (TM4)], D[=Asp883 (TM6)]. Two of the non-catalytic Ca^{2+}-binding sites at the C-terminus of the protein are shown by loops. (**B**) Effect of calmodulin on the PMCA pump. The receptor site for the calmodulin-binding domain blocks the access for Ca^{2+} and CaM. Binding calmodulin displaces the peptide, freeing the access to the active sites

confirmed the presence of these four isoforms in different organisms. Protein-sequencing data has validated the existence of the human PMCA1 and PMCA4 isoforms as proteins [12, 14, 16]. Sequence comparisons [17] have shown that the coding sequences of the PMCA isoforms are highly conserved in the region spanning from the 20–30th N-terminal amino acid to the C-terminally located calmodulin-binding region (at about the 1100th amino acid), but are highly divergent at the C- and N-termini.

2.3 Identification of Functional Domains

Hydropathy analysis of the pump [14] suggested the presence of 10 transmembrane domains connected on the external side by five short loops (Fig. 2). The pump protrudes into the cytoplasm with three main units: The first one between transmembrane domains 2 and 3 contains the acidic phospholipid interacting site [18], the second one between transmembrane domains 4 and 5, the catalytic site [12], and the third and C-terminal one, after transmembrane domain 10, are sites involved in the regulation of the activity of the pump. The latter encompass the calmodulin-binding domain [12], the substrate domains for protein kinases A and C [19, 20], and a domain which contains high-affinity Ca^{2+}-binding sites [21]. The N-terminal 80–90 residues located in the cytosol are essential for the activity and may play an important role in the subcellular localization of the pump [22]. Experimental support for the topological model has now come from antibodies binding work coupled to proteolysis and peptide analysis of pump fragments [23] and to genetic deletions of portions of the pump [24].

3 Regulation of the Activity the Pump

In the basal state the pump has low Ca^{2+} affinity ($K_m > 10$ μM). A number of regulators increase the Ca^{2+} affinity to a K_m as low as 0.2 μM. The most important activator of the pump, calmodulin, changes its affinity to Ca^{2+} from a K_m of 10–50 μM to 0.2–0.3 μM [23]. The pump can also be stimulated by acidic phospholipids [25, 26].

As for many other calmodulin-dependent enzymes, the pump becomes activated when calmodulin displaces an inhibitory peptide from the catalytic site (Fig. 2). In the case of the PMCA the inhibitory peptide was shown to be 28 amino acids long [27]. Two binding sites were identified on the pump for this peptide. One site (residues 537–544) [28] is located in the large cytoplasmic loop between the site of aspartic acid phosphorylation and that of ATP binding. The other site (residues 206–271) [29] is located in the cytosolic loop between transmembrane domains 2 and 3 (Fig. 2).

The PMCA1 pump isoform can be phosphorylated and stimulated by the cAMP-dependent protein kinase [19, 30, 31] and protein kinase C [20] while PMCA4 has been shown to be phosphorylated at Tyr 1176 causing a reduction in its activity [32]. Recently, phosphorylation of the PMCA3 and PMCA2 pump isoforms has been observed in the brain [33]. Despite efforts by many laboratories, the physiological significance of the phosphorylation of the PMCA pump is still a matter of debate.

Experiments with synthetic peptides have suggested that phosphorylation by protein kinase C (at Thr 1102) reduces the affinity of the calmodulin-binding domain with the binding ("receptor") site in the pump [34]. Further work has indicated that protein kinase C also mediates phosphorylation at different sites to Thr 1102 [35]. Since in the PMCA pump isoforms the sequences recognized by protein kinases are not conserved and are subjected to alternative splicing, phosphorylation is expected to have different effects on the activity of pump isoforms.

4
Structure of the PMCA Genes and Their Chromosomal Localization

Structures of the genes of the rat PMCA3 and human PMCA1 pumps have been published [36, 37]. Both genes span a large genomic region of at least 90 kbp. The structures of the rat and human genes are very similar and all the intron-exon boundaries coincide, the only exception being exons 22–23 in PMCA3. The lack of a TATA box and the high GC content in the putative promoter region of the PMCA1 is consistent with the housekeeping role of this plasma membrane Ca^{2+} ATPase isoform. Information on the genomic structure is not yet available for the PMCA2 and 4 isoforms, while the promoter region of the PMCA3 has not yet been identified.

The human PMCA1 and PMCA4 genes have been located on chromosomes 12 (q21–q23) and 1 (q25–q37), respectively [38]. The human PMCA2 and PMCA3 genes are located on human chromosomes 3p23–p25 and Xq28, respectively [39]. Recent sequencing data indicate that a portion of the PMCA 2 gene 3′-end is located on the bacmid spanning 3q25–26 region of the human chromosome (EMBL Data base, AC008116). The region corresponding to the human chromosome 3p25–26 is homologous in the mouse to a region of chomosome 6 that is responsible for the deaf-waddler (dfw) mutation. Mice carrying this mutation are deaf [40]. Genetic analysis demonstrated that a mutaton in the PMCA2 gene was responsible for this phenotype [40].

5
Isoforms of the PMCA Pump

5.1
Isoforms Generated by Alternative Splicing

Considerable differences at the primary sequence level have been detected in the regions subjected to alternative splicing of the primary transcripts of the pumps. The two sites at which alternative splicing occurs are located in the cytosolic loop between transmembrane domains 2 and 3 (site A) and within the calmodulin-binding domain (site C). In the case of rat PMCA3, alternative splicing at site C generated up to seven different isoforms [41]. The exons that can be spliced at site C modify substantially the sequence of the calmodulin-binding domain (Fig. 3) [42].

The biochemical properties of the PMCA pump proteins were studied after their overexpression in eukaryotic systems. Some of the properties of the PMCA isoforms

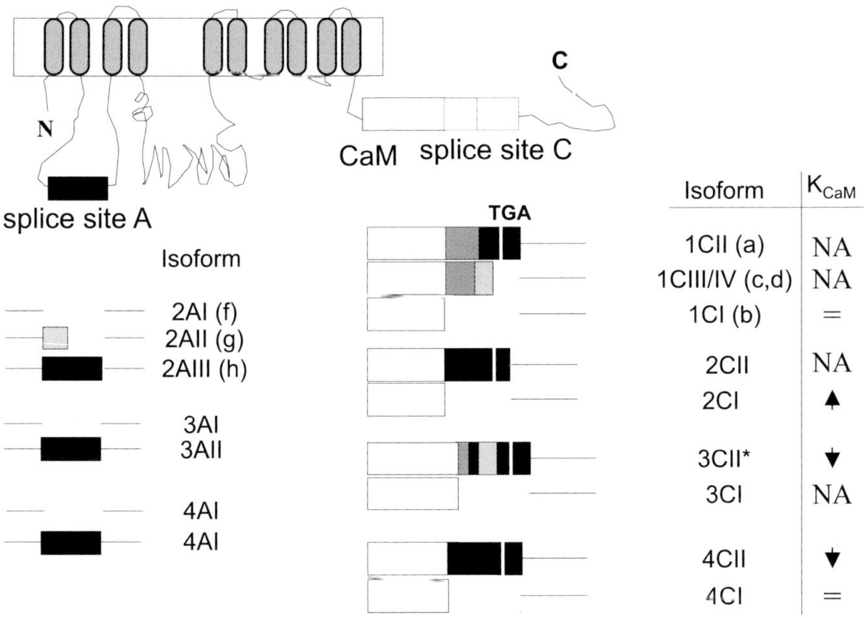

Fig. 3. Alternative splicing of the PMCA. A topological model of the PMCA (*gray barrels* represent transmembrane domains) and the location of the two alternative splicing sites are shown at the top of the panel. In the lower portion of the panel, some of the possible (most common) splicing products are indicated. The *black and gray boxes* represent the sequences involved in splicing and their approximate size. The *box shadings* indicate that the regions are encoded by different exons or that they are the products of independent internal splicing. The name of the PMCA isoforms is according to Carafoli [77]. In the parentheses another nomenclature is given that is used to name PMCA isoforms. In the case of the C-splice variants the effect on the affinity toward calmodulin is indicated: *NA*: not available; = the affinity identical to that of PMCA4CI; ↓ the affinity is weaker or ↑ stronger than that of PMCA4CI. (Guerini et al. in preparation)

Table 1. Properties of the PMCA isoforms

	PMCA1CI	PMCA1CII	PMCA2CI	PMCA3CII	PMCA4CI	PMCA4CII
Tissue distribution	ubiquitous	restricted	restricted	restricted	ubiquitous	restricted
Tissue with high expression	many	brain/muscle	brain	brain	many	brain/muscle
Level of expression	high	high	high	medium/low	medium	medium
Calmodulin affinity Kd (nM)[a]	40–50	NA (low)	5–10	300–500[b]	40–50	500
Acidic phospholipid activation	NA	NA	yes	NA	yes	NA

NA data not available.

[a] The values for calmodulin affinity were obtained on recombinant proteins expressed in Sf9 cells [13, 45]

[b] Guerini et al. in preparation.

are summarized in Table 1. Notably, the most significant difference between the PMCA isoforms is the effect of calmodulin on their activity. The PMCA2 pump has the highest affinity to calmodulin [43], while alternative splicing at site C (PMCA4CII) reduced the affinity to calmodulin by 20 to 50 times [44, 45]. The PMCA4CII pump, the isoform that showed a 20-fold reduction in affinity for calmodulin, has higher basal activity.

Alternative splicing at site A did not seem to affect the activity of the pump [43]. A detailed analysis of the region located C-terminally to the calmodulin-binding regions indicated the importance of this region for the auto-inhibition of the pump [46]. Alternative splicing was observed at two additional sites called B and D [17] but they turned out to be a cloning artifact and/or to generate inactive protein. For example, the PMCA4BICI pump, a splice variant that lacked most of the region corresponding to the 9th transmembrane domain, found in large amounts in liver [47], was inactive when overexpressed and purified from insect cells [45].

5.2
PMCA Isoforms Tissue Distribution and the Control of Their Expression

Large differences have been observed in the tissue distribution of the PMCA isoforms. The mRNAs of rat and human PMCA1 and PMCA4 pumps have broad tissue distribution, whereas those of PMCA2 and PMCA3 pumps have only been detected in brain and heart [41, 48, 49]. Some of the alternatively spliced isoforms of the four PMCA genes also showed tissue-specific distribution [50]. In situ hybridization on rat brain has shown substantial differences in the regional distribution of the mRNAs of isoforms 1, 2, 3 and 4 [48, 51]. In particular, high amounts of PMCA1 mRNA have been detected in the hippocampus, PMCA2 transcripts were especially abundant in Purkinje cells of the cerebellum and finally, PMCA3 mRNA was high in the habenula and the choroid plexuses. The PMCA4 transcripts were very abundant in the Piriform cortex, but absent in Purkinje cells and the habenula [51]. Isoform-specific antibodies have confirmed observations at the mRNA level [52, 53]. It has become clear that the brain contains amounts of PMCA2 and PMCA3 proteins comparable to those of PMCA1 and PMCA4. In addition, brain is the tissue with the highest amount of plasma membrane pump protein (up to 10 times more than in non-excitable cells). These observations suggest that the expression of the PMCA2 and PMCA3 isoforms is controlled by different mechanisms than those controlling the two other PMCA isoforms.

Cerebellar granule cells have been frequently used to study the development of nervous cells and have evolved as a very useful tool to study the expression of the PMCA pump. A marked increase in the total PMCA pump protein was detected during the in vitro maturation of these cells in the presence of 25 mM KCl [54]. While the PMCA1CI splice variant remained constant during the maturation process, the CII variant increased markedly. The expression of PMCA2 and PMCA3 also increased dramatically during the maturation of the granule cells [54]. The up-regulation of the pump isoforms was dependent on the chronic elevation of the cytosolic Ca^{2+}, which was the result of a partial depolarization of the plasma membrane. The effect was specific for the PMCA pump, since no changes were seen for the SERCA pump.

The up-regulated isoforms (PMCA2, PMCA3, PMCA1CII) are those typical of adult brain tissue [48, 49, 52]. The most significant increase was that of PMCA2: In mature neurons this isoform represented more than 30–40% of the total pump protein. The large increase in the amount of pump protein was likely a reflection of the stimulation of the transcription of the genes (Fig. 4). Other as yet unidentified post-transcriptional and/or post-translational Ca^{2+}-sensitive mechanisms may also have played a role.

The functional consequences of the PMCA3 pump increase cannot be assessed, since the biochemical characterization of this pump isoform has not been described. As for the functional properties of isoform 1CII, which also became significantly up-regulated, they could be tentatively derived from those of the homologous 4CII isoform (Fig. 3). The splicing process results, in the case of PMCA4, in a pump that has weaker binding affinity to calmodulin, but also higher basal (i.e., calmodulin-independent) ATPase activity. The up-regulation of PMCA1CII correlated well with the down-regulation of PMCA4CII in the granule cells. The significance of this process is not known but may depend on the different subcellular localization of the

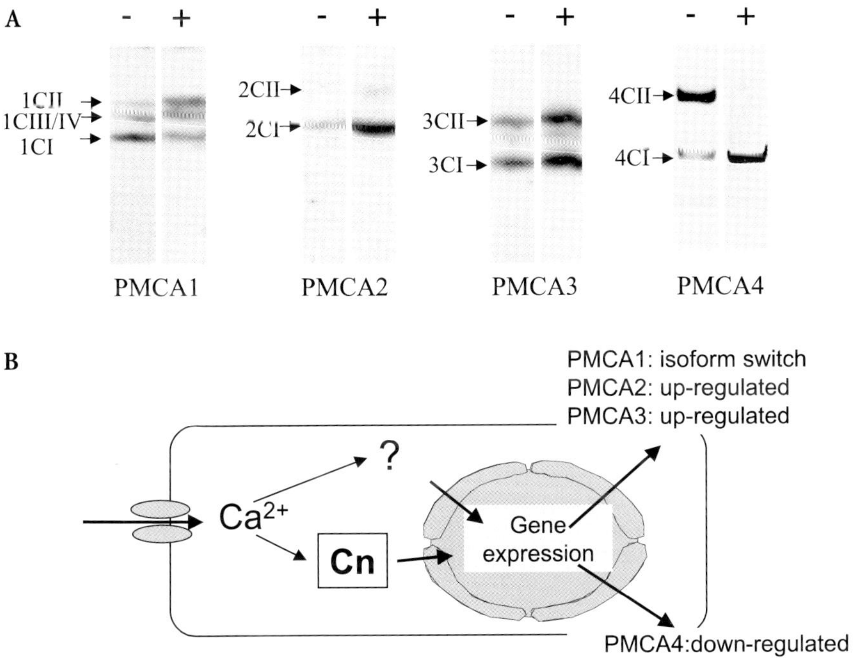

Fig. 4. KCl dependence of the expression of the PMCA pumps. (**A**) RT-PCR on transcripts from cerebellar granule cells cultured in the presence of 5.3 (–) or 25 (+) mM KCl for 5 d using isoform-specific oligonucleotides. The oligonucleotides amplify a portion (200–300 nt) of mRNA surrounding the splice site C. (**B**) Summary of the effect of Ca^{2+} and calcineurin (Cn) on the expression of PMCA isoforms in granule neurons. For details see [54]. The difference in the intensity of the PMCA4CI bands (– and +) is an artifact due to competition of the two products (CI and CII) for the same oligonucleotides. This difference disappears if PMCA4CI alone is analyzed, i.e. no up- or down-regulation of PMCA4CI is observed

two isoforms. During maturation neurons undergo a large rearrangement of their shape and, in particular, they generate a dense network of cell-to-cell connections (synapses). In the case of PMCA4, but not for other PMCA isoforms, association with PSD-95, a protein responsible for post-synaptic localization of neuronal proteins, was demonstrated [55]. This finding suggests that PMCA isoforms may have different subcellular localization: down-regulation or up-regulation of a certain PMCA isoform may result in changes in the spatial distribution of the pump activity. As for the PMCA2 pump, it has the highest affinity for calmodulin of all PMCA pumps. An increase in the PMCA2 pump may thus be expected to provide granule cells with higher sensitivity to lower concentrations of Ca^{2+}-calmodulin. Since the Ca^{2+}-calmodulin-PMCA complex is the active state of the pump, an increased affinity for calmodulin should result in an improved response of the pump to lower concentrations of Ca^{2+}.

As mentioned above, the PMCA4CII isoform behaves differently to the other pump isoforms: depolarization induces its rapid disappearance from the cell membrane. In the case of PMCA4CII this process is regulated by Ca^{2+} and by calcineurin, while the up-regulation of the other PMCA isoforms is dependent on Ca^{2+}, but not affected by calcineurin (Fig. 4). Culturing the neurons in the presence of FK506, an immunosuppressant that specifically inhibits calcineurin, blocks completely the depolarization and Ca^{2+} effects on PMCA4CII, while the up-regulation of PMCA1CII, PMCA2 and PMCA3 is unaffected. It is not yet known if Ca^{2+} regulates the expression of PMCA1CII, 2 and 3 directly, by a factor like DREAM (downstream regulatory element antagonist modulator) [56] or by a CaM-kinase pathway.

During nerve growth factor (NGF) mediated differentiation of PC-12 cells to neurons, changes in the patterns of alternative splicing of the four pump isoforms were observed [4]. In neuroblastoma cells, alternative splicing at site A of the PMCA2 pump isoform has been shown to be dependent on KCl-mediated depolarization [57].

6 Membrane Targeting of the PMCA Pump

Attempts to reproduce the effects of proteolysis by expressing pumps lacking portions of the C-terminal domain in COS cells unexpectedly led to the discovery of a retention signal for ER in the PMCA molecule. The cells expressing the full-length ATPase and $PMCA_{\Delta 139}$ had a pattern characteristic of cell surface expression. By contrast, cells expressing $PMCA_{\Delta 118}$ showed retention in the ER [58]. Interestingly, in the samples containing the $PMCA_{\Delta 118}$, a polypeptide of about 82 kDa was present which was never detected in membrane preparations of other expressed polypeptides and which was tentatively assumed to be a degradation product of the $PMCA_{\Delta 118}$. Analysis of $PMCA_{\Delta 118}$ containing membranes with antibodies revealed that the 82 kDa degradation product was more abundant than the 125 kDa polypeptide (Fig. 5).

The retention in the ER was apparently associated with the 21 amino acid sequence between Glu1067 and Arg1087. This sequence is highly acidic and contains residues conserved in all PMCA isoforms. The mutation of only one Glu to Gln was

A

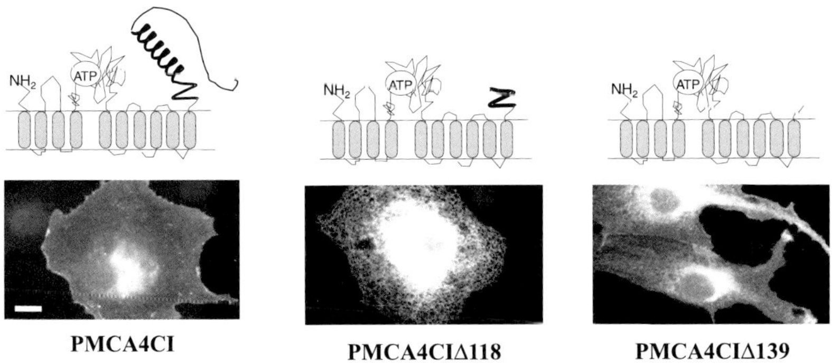

B PMCA4CIΔ118

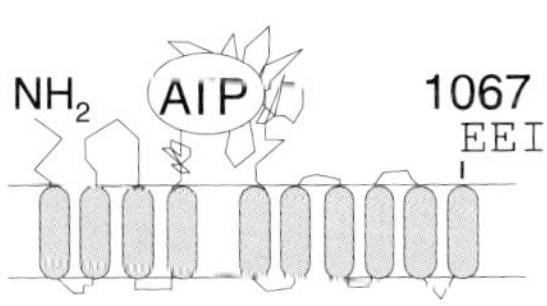

	Sub loc	Degr
EEITKDAEGLDEIDHAEMELR	ER	Yes
L A A A	PM	No
Q Q Q	PM	No
Q Q	PM	No
Q N	PM	No
Q	PM	No

Fig. 5. A hidden retention-degradation signal in the PMCA pump. (**A**) Immunocytochemistry of COS cells expressing PMCA4CI and two C-terminally truncated mutants: Δ118 and Δ139 lacking 118 and 139 amino acid residues at their C-terminus, respectively. The Δ118 mutant shows a typical endoplasmic reticular staining, the other the staining for protein delivered to the plasma membrane. At the top of each picture a schematic representation of the topology of the mutants is given. All the mutants were active. (**B**) Summary of the single amino acid mutations (in the peptide Glu1067–Arg1087) that reverted the endoplasmic reticulum retention and degradation of the Δ118 mutant. *Sub loc* subcellular localization; *ER* endoplasmic reticulum; *PM* plasma membrane; *Degr* generation of the 80 kDa proteolytic fragment. For more details see [58]

sufficient to restore the delivery to the plasma membrane and prevent the degradation of the truncated pump. The properties of the Glu1067–Arg1087 sequence are very similar to those of the Pro1134–Asp1157 sequence: both contain a preponderance of acidic amino acids and the calculated PEST score for both sequences is very high [59, 60]. They both bind Ca^{2+} with high affinity [34]. Nevertheless, substitution of the Glu1067–Arg1087 sequence in the truncated mutants with the Pro1134–Asp1157 sequence failed to cause retention in the ER or to generate proteolytic products [58]. The selective ER retention and the degradation of the truncated PMCA pump could have evolved as a defense mechanism to prevent the release of a prematurely terminated protein with undesirable functional properties. The shortened version of the pump is fully active in the absence of calmodulin [61]. If a fully active, calmodulin independent, truncated protein were delivered to the plasma membrane, it would pump calcium out of the cell in an uncontrolled manner, very likely resulting in cell damage.

The SERCA and PMCA pumps differ in their subcellular localization. Since the SERCA Ca^{2+} pump is only present in the ER, the existence of signal sequences for its retention in this organelle were predicted. To explore the matter, chimeric molecules of the PMCA and SERCA pumps, in which the N-terminal regions had been interchanged, were constructed and expressed in COS cells. In the case of chimera E, a chimeric SERCA pump that carries the region encompassing the first two transmembrane domains of the PMCA pump, the majority of the expressing cells showed a reticular staining, but a relevant fraction of the transfected cells showed expression in the plasma membrane [62]. These experiments suggest that the N-terminal portion of the SERCA pump contains signals for efficient retention in the endoplasmic reticulum (Fig. 6).

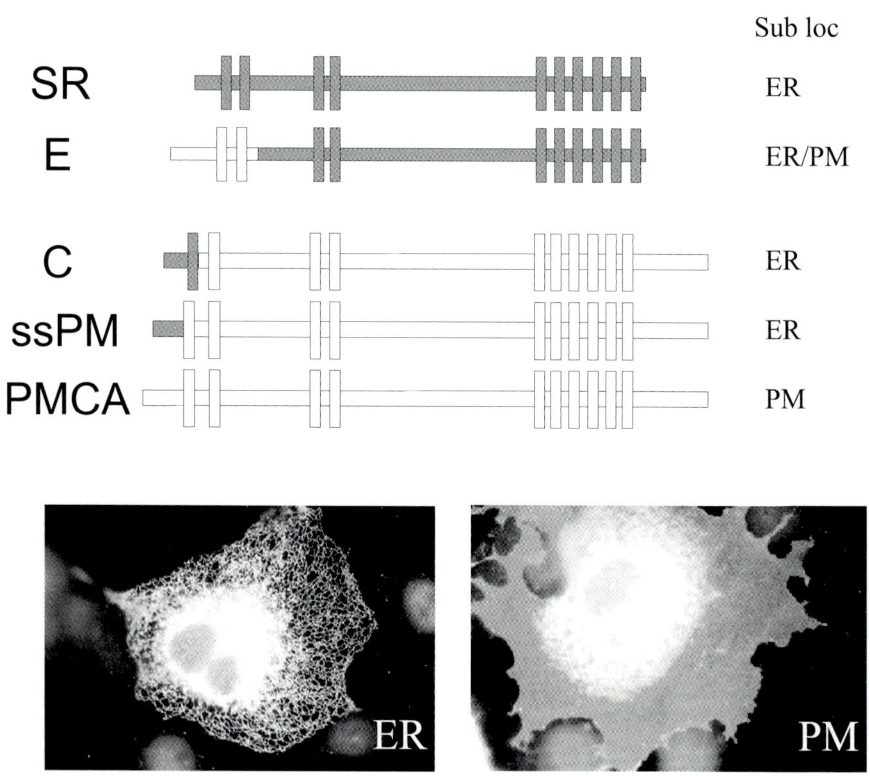

Fig. 6. Role of the N-terminal domain in the retention and the activity of the SERCA and PMCA pumps. Schematic representation of chimeric proteins (SERCA in *gray*, PMCA in *white*) used to study the N-terminal domain of both pumps. All chimera (E, C and ssPM) were inactive or had lost most of their catalytic activity [62, 63]. In the case of chimera E a significant portion of the cells (10%) expressing the recombinant protein showed staining of the plasma membrane (PM), the remainder endoplasmic reticular staining (ER). *Sub loc* subcellular localization

Further studies with reverse chimeras (C and ssSRM) confirmed that the N-terminal portion of the SERCA protein contained an ER retention signal [63]. These experiments suggest that sequences located in the N-terminal domain are critical for the activity of the pump (Fig. 2).

7 Path of Ca^{2+} Through the Plasma Membrane Ca^{2+} ATPase

Site-directed mutagenesis has been used to study the role of some of the amino acids present in the PMCA pumps. Amino acids in the putative ATP-binding region, which are predicted to be crucial for the correct folding of the ATP-binding domain of P-type pumps [64], were mutated. The mutation of one of them (Asp672) caused a major loss of activity, which resulted from the slower E_1-E_2 conformational transition [65].

Mutations were inserted in the transmembrane domains (TM) 4, 5, 6 and 8 (Fig. 7). Those inserted at the position of Asn879 and Asp883 (both in TM6), of Glu423 (TM4), and of Gln971 (TM8) [66, 67] abolished the transport of Ca^{2+} and the formation of the phosphoenzyme intermediate from ATP by the pump. Mutated pumps, in which Glu423 (TM4) or Asp883 (TM6) were substituted, formed high amounts of phosphorylated intermediate from phosphate, much higher than the wild-type pump (Fig. 7) [67]. Differences in the cellular localization of the mutants were also observed; while Asn879 and Gln971 mutants were retained in the ER, the other two were correctly delivered to the plasma membrane. This indicates that the mutations of Asn879 and Gln971 prevent the pump from acquiring the conformation required for proper delivery to the plasma membrane. In contrast the phenotype of the Glu423 and Asp883 mutations was consistent with their role in the translocation of Ca^{2+}. In particular the capacity of the mutants to form high amounts of phosphorylated intermediate from phosphate indicated that only the Ca^{2+}-dependent portion of the catalytic cycle was affected. The PMCA pump lacks two residues that have been proposed to donate oxygen atoms for the second Ca^{2+}-binding site in the SERCA pump, i.e. Glu771 (TM5) and Thr799 (TM6) [68, 69].

Mutations of Met882 (TM6), corresponding to the position of Thr 799 (TM6) in the SERCA pump, had no obvious effects on the activity of the PMCA pump [66]. Mutation of Ala854, the homologue of Glu771 of the SERCA pump, aimed at introducing charged or polar residues in transmembrane domain 5 had unexpected consequences: The mutated pump was active when the Ala residue was mutated to a Glu or Gln residue, but lost its activity when changed to an Asp. All Ala854 mutants (active or not) were retained in the ER. After purification to homogeneity, the Ala854Glu and Ala854Gln mutants still had 30 to 40% of the wild-type pump specific activity. Kinetic analysis indicated that these two mutants had acquired properties typical for the SERCA pump (Zecca et al. in preparation). These experiments agree with the presence in the PMCA of two ligands for Ca^{2+} [Glu423 (TM4) and Asp883 (TM6)]. Insertion of a charged (polar) amino acid at TM5 (at the position of Ala854) made the PMCA more similar to the SERCA pump. This is consistent with the idea that the PMCA pump transports one Ca^{2+} per reaction cycle, and that the Glu in TM5 is essential for the transport of the second Ca^{2+} in the SERCA pump (Fig. 7).

	ATPase	Pin-ATP	Pin-Pi	Ca^{2+}	Local
Glu423Ala	-	-	+	+	PM
Ala854Glu	+	+	+	+	ER
Asn879Ala	-	-	-	-	ER
Asp883Ala	-	-	+	+	PM
Gln971Ala	-	-	-	-	ER

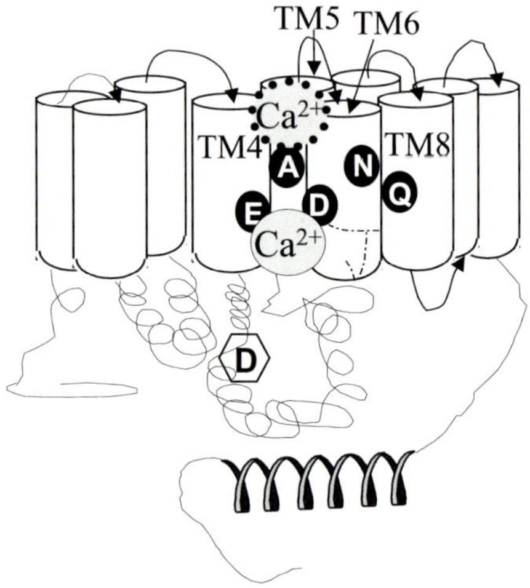

Fig. 7. Amino acids of the PMCA involved in the translocation of Ca^{2+}. *Top*: Summary of the properties of pumps carrying mutations in five amino acid residues in the transmembrane domains 4, 5, 6, and 8. The homologous residues were suggested to be involved in the translocation of Ca^{2+} in the SERCA pump [78]. In the case of the PMCA only Glu423 and Asp883 were shown to be involved (see text). The following properties of the PMCA pump were analyzed: the Ca^{2+}-dependent ATPase activity (ATPase), the formation of the phosphoenzyme intermediate from ATP (P_{in}-ATP) or from phosphate (P_{in}-Pi), the sensitivity to Ca^{2+} and the subcellular localization (Local). The subcellular localization was determined by immunocytochemistry after the transient expression in COS cells. *Bottom*: Model showing the location of the five amino acids (*black dots*) that where mutated and whose properties are summarized in the top portion of the panel. The location of the Asp residue in the catalytic domain (D = D465, see also Fig. 2) is also indicated. The single high-affinity Ca^{2+} translocation site of the PMCA pump is indicated by the *white ellipsoid*: It is flanked by Glu423 and Asp883. The *dotted ellipsoid* shows the speculative location of a second Ca^{2+}-binding (translocation) site, which may arise when Ala854 is mutated to a Glu or a Gln (Zecca et al. in preparation). This is the proposed location of the second Ca^{2+} translocation site in the SERCA pump [68]

Equally puzzling was the observation that a single substitution would lead to the retention of an active PMCA mutant in the ER. A possible interpretation is that the insertion of a charged (polar) residue with the correct size in transmembrane domain 5 of the PMCA pump translated in the generation of a "conformational" retention signal. This would suggest that the retention of the SERCA pump in the ER is not only dependent on linear retention signals, but is also mediated by spatial information associated with the packaging of the transmembrane domains.

8 PMCA in Disease

A limited number of disease conditions have been traced back to alterations in the activity of the PMCA pump. These include uncontrolled diabetes [70–72] and sickle cell disease [73–75]. A decreased V_{max} of the PMCA pump in sickle cell anaemia has been repeatedly reported [73]. In one of the studies, which was unfortunately performed before the pump had been cloned, the enzyme was purified from sickled erythrocytes, and found to behave normally [75]. Decreased Ca^{2+} ATPase activity has been reported in the erythrocytes of uncontrolled diabetics [70, 71] and of experimentally diabetic rats [72].

Recently, the mutation responsible for deaf-waddler (dfw) mice, which causes deafness and gross locomotor problems, was mapped at the position where the PMCA2 gene was found. The mutation resulted in a dramatic decrease in the amount of PMCA2 mRNA. Immunohystochemical analysis of the inner-ear cells of dfw mice demonstrated a strongly reduced signal for the PMCA pump protein [40]. Knock-out mice carrying a deleted PMCA2 gene had a defect similar to that observed in dfw mice [76]. These experiments demonstrated that PMCA2 is essential for the proper function of inner-ear cells signal transduction. Inactivation of the PMCA2 gene likely leads to the continuous activation of the inner-ear sensory cells (because of a sustained high level of cytosolic Ca^{2+}) and to deafness.

9 Summary

Combinations of biochemical and molecular biological approaches have been used to determine the functional domains of the plasma membrane Ca^{2+} pump. In particular, recent efforts show that the PMCA isoforms have differences in their affinity to calmodulin and in their basal (calmodulin-independent) activity. The expression of some of the PMCA isoforms is in neurons, at least, controlled by Ca^{2+} and in one case by the protein phosphatase calcineurin. This is to some extent surprising because the PMCA pump has always been considered a housekeeping enzyme and its expression was therefore not expected to undergo tissue and development regulation. Site-directed mutagenesis has provided information on amino acids that may be involved in the translocation of Ca^{2+} and, even more strikingly, indicates that one of them might be the residue that is critical for the differences between the PMCA and the SERCA pumps. Many important issues at the molecular, cell biological and biochemical level still have to be addressed. The regulation of the expression of the pumps should be further investigated in particular to find out which, beside Ca^{2+} and calcineurin, are the other players involved. It will also be important to translate in a cellular (in vivo) system the biochemical (in vitro) observations: how does the switch between a set of one isoform to another affect Ca^{2+} homeostasis and how does it affect the kinetics of the intracellular Ca^{2+} transients. The dwf and the PMCA2 knock-out mice are new valuable tools to study the role of PMCA2 in the brain. Finally, the possibility to produce pure recombinant PMCA isoforms may stimulate efforts to solve a crystal structure of this protein.

Acknowledgements. The author would like to thank Prof. Dr J. Krebs and Prof. E. Carafoli for stimulating discussions and Dr Adriana Sburlati for critically reading this manuscript. The work was made possible by the financial support of the Swiss National Science Foundation (Grant No. 31-30858.91).

References

1. Carafoli E (1987) Ann Rev Biochem 56:395
2. Philipson KD, Nicoll DA (1992) Curr Opin Cell Biol 4:678
3. Pedersen PL, Carafoli E (1987) Trends Biochem Sci 12:186
4. Hammes A, Oberdorf S, Strehler EE, Stauffer T, Carafoli E, Vetter H, Neyses L (1994) FASEB J 8:428
5. Pedersen PL, Carafoli E (1987) Trends in Biochem Sci 12:146
6. Jencks WP (1992) Ann NY Acad Sci 671:49
7. Herscher CJ, Rega AF (1996) Biochemistry 35:14917
8. Guerini D, Carafoli E (1999) The calcium pumps. In: Carafoli E, Klee C (eds) Calcium as a cellular regulator. Oxford University Press, New York, pp 249–278
9. Schatzmann HJ (1966) Experientia 22:364
10. Niggli V, Penniston JT, Carafoli E (1979) J Biol Chem 254:9955
11. Carafoli E (1991) Physiol Rev 71:129
12. James P, Maeda M, Fisher R, Verma AK, Krebs J, Penniston JT, Carafoli E (1988) J Biol Chem 263:2905
13. Shull GE, Greeb J (1988) J Biol Chem 263:8646
14. Verma AK, Filoteo AG, Standford DR, Wiebenm ED, Penniston JT, Strehler EE, Fisher R, Heim R, Vogel G, Mathews S, Strehler-Page M-A, James P, Vorherr T, Krebs J, Carafoli E (1988) J Biol Chem 263:14159
15. Greeb J, Shull GE (1989) J Biol Chem 264:18569
16. Strehler EE, James P, Fisher R, Heim R, Vorherr T, Filoteo AG, Penniston J, Carafoli E (1990) J Biol Chem 265:2835
17. Strehler EE (1991) J Membrane Biol 120:1
18. Zvaritch E, James P, Vorherr T, Falchetto R, Modyanow N, Carafoli E (1990) Biochemistry 29:8070
19. James P, Pruschy M, Vorherr T, Penniston JT, Carafoli E (1989) Biochemistry 28:4253
20. Wang KKW, Wright LC, Machan CL, Allen BG, Conigrave AD, Roufogalis BD (1991) J Biol Chem 266:9078
21. Hofmann F, James P, Vorherr T, Carafoli E (1993) J Biol Chem 268:10252
22. Guerini D, Carafoli E (1996) Bioscience Reports 16:129
23. Carafoli E (1992) J Biol Chem 267:2115
24. Adamo H, Caride AJ, Penniston JT (1992) J Biol Chem 267:14244
25. Niggli V, Adunyah ES, Carafoli E (1981) J Biol Chem 256:8588
26. Brodin P, Falchetto R, Vorherr T, Carafoli E (1992) Eur J Biochem 204:939
27. Vorherr T, James P, Krebs J, Enyedi A, McCormick DJ, Penniston JT, Carafoli E (1990) Biochemistry 29:355
28. Falchetto R, Vorherr T, Brunner J, Carafoli E (1991) J Biol Chem 266:2930
29. Falchetto R, Vorherr T, Carafoli E (1992) Protein Sci 1:1613
30. Caroni P, Carafoli E (1981) J Biol Chem 256:9371
31. Neyses L, Reinlieb L, Carafoli E (1985) J Biol Chem 260:10283
32. Dean W, Chen D, Brandt P, Vanaman TC (1997) J Biol Chem 272:15113
33. Zylinska L, Guerini D, Gromadzinska E, Lachowicz L (1998) Biochim Biophys Acta 1448:99
34. Hofmann F, Anagli J, Carafoli E, Vorherr T (1994) J Biol Chem 269:24298
35. Enyedi A, Verma AK, Filoteo AG, Penniston JT (1996) J Biol Chem 271:32461
36. Burk SE, Shull GE (1992) J Biol Chem 267:19683
37. Hilfiker E, Strehler M-A, Carafoli E, Strehler EE (1993) J Biol Chem 268:19717
38. Olson S, Wang MG, Carafoli E, Strehler EE, McBride OW (1991) Genomics 9:629

39. Wang GM, Huafang Y, Hilfiker H, Carafoli E, Strehler EE, McBride OW (1994) Cytogenet Cell Gen 67:41
40. Street VA, McKee-Johnson JW, Fonseca RC, Tempel BL, Noben-Trauth K (1998) Nat Genet 19:390
41. Keeton TP, Burk SE, Shull GE (1993) J Biol Chem 268:2740
42. Carafoli E, Guerini D (1993) Trends Cardiovasc Med 3:177
43. Hilfiker H, Guerini D, Carafoli E (1994) J Biol Chem 269:26178
44. Enyedi A, Verma AK, Heim R, Adamo HP, Filoteo AG, Strehler EE, Penniston JP (1994) J Biol Chem 269:41
45. Seiz-Preianò B, Guerini D, Carafoli E (1996) Biochemistry 35:7946
46. Verma AK, Enyedi A, Filoteo AG, Strehler EE, Penniston JT (1996) J Biol Chem 271:3714
47. Howard A, Legon S, Walters JR (1993) Am J Physiol 265:G917
48. Stahl WL, Eakin TJ, Owens JWM, Breininger JF, Filuk PE, Anderson WR (1992) Mol Brain Res 16:223
49. Stauffer T, Hilfiker H, Carafoli E, Strehler EE (1993) J Biol Chem 268:25993
50. Strehler EE, Strehler-Page M-A, Vogel G, Carafoli E (1989) Proc Natl Acad Sci USA 86:6908
51. Stahl WL, Keeton TP, Eakin TJ (1994) Neurosci Lett 178:267
52. Stauffer T, Guerini D, Carafoli E (1995) J Biol Chem 270:12184
53. Stauffer TP, Guerini D, Celio MR, Carafoli E (1997) Brain Res 748:21
54. Guerini D, Garcia Martin E, Gerber A, Volbracht C, Leist M, Merino CG, Carafoli E (1999) J Biol Chem 274:1667
55. Kim E, DeMarco SJ, Marfatia SM, Chishti AH, Sheng M, Strehler EE (1998) J Biol Chem 273:1591
56. Carrion AM, Link WA, Ledo F, Mellstrom B, Naranjo JR (1999) Nature 398:80
57. Zacharias DA, Strehler EE (1996) Curr Biol 6:1642
58. Zvaritch E, Vellani F, Guerini D, Carafoli E (1995) J Biol Chem 270:2679
59. Rechsteiner M (1988) Adv Enzyme Regul 27:135
60. Wang K, Villalobo A, Roufogalis BD (1989) Biochem J 262:693
61. James P, Vorherr T, Krebs J, Morelli A, Castello G, McCormick DJ, Penniston JT, De Flora A, Carafoli E (1989) J Biol Chem 264:8289
62. Foletti D, Guerini D, Carafoli E (1995) FASEB J 9:670
63. Guerini D, Garcia Martin E, Zecca A, Guidi F, Carafoli E (1998) Acta Physiol Scand Suppl 643:265
64. Taylor WP, Green MN (1989) Eur J Biochem 179:241
65. Adamo HP, Filoteo AG, Enyedi A, Penniston JT (1995) J Biol Chem 270:30111
66. Adebayo AO, Enyedi A, Verma AK, Filoteo AG, Penniston JT (1995) J Biol Chem 270:27812
67. Guerini D, Foletti D, Vellani F, Carafoli E (1996) Biochemistry 35:3290
68. Andersen JP (1995) Bioscience Reports 15:243
69. Chen L, Sumbilla C, Lewis D, Zhong L, Strock C, Kirtley ME, Inesi G (1996) J Biol Chem 271:10745
70. Schaefer W, Priessen J, Mannhold R, Gries AF (1987) Klin Wochenschr 65:17
71. Gonzalez-Flecha FL, Bermunez M, Cedola NV, Gagliardino JJ, Rossi JPFC (1990) Diabetes 39:707
72. Reddi AS, Dasmahapatra A, Jyothirmayi GN, Jayasundaramma B (1992) Am J Hypertens 5 (121 pt 1):863
73. Bookchin RM, Lew VL (1980) Nature 284:561
74. Gopinath RM, Vincenzi FF (1981) Am J Hematol 7:303
75. Niggli V, Adunyah ES, Cameron BF, Bababunmi EA, Carafoli E (1982) Cell Calcium 3:131
76. Kozel PJ, Friedman RA, Erway LC, Yamoah EN, Liu LH, Riddle T, Duffy JJ, Doetschman T, Miller ML, Cardell EL, Shull GE (1998) J Biol Chem 273:18693
77. Carafoli E (1994) FASEB J 8:993
78. Clarke DM, Loo TW, Inesi G, MacLennan DH (1989) Nature 339:476

The Na^+/Ca^{2+} Exchanger: Structural Aspects, Function and Regulation

Nadia Gabellini, Alessandra Zatti, Ernesto Carafoli*

Department of Biological Chemistry, University of Padova, Via G. Colombo 3, I-35121 Padova, Italy
E-mail: carafoli@civ.bio.unipd.it

The Na^+/Ca^{2+} exchanger of the plasma membrane is a reversible transporter that couples the translocation of 3 Na^+ against 1 Ca^{2+}. Its activity is particularly important to excitable tissues. Three homologous exchanger gene products (NCX1, 2, 3) are differently expressed in heart, brain and kidney. Hydropathy analysis has indicated 11 hydrophobic segments with highly conserved regions that have been suggested to play a role in ion transport (α-repeats). The number of transmembrane domains has recently been revised down from 11 to 9 in an alternative topography model. The NCX1 gene includes an unusually large exon (2) that encodes about 2/3 of the protein (108 kDa), i.e. a partially active protein of ~70 kDa. The mRNA levels of the exchanger are controlled by tissue-specific promoters. In some cell types (e.g. neurons) they are modulated by Ca^{2+}: in the case of NCX2, the Ca^{2+} modulation is mediated by the calmodulin-dependent protein phosphatase calcineurin. The regulatory mechanisms of autoinhibition and activation by Ca^{2+} and ATP are linked to domains in the large central cytosolic loop of the protein.

Keywords. Na^+/Ca^{2+} exchanger, 70 kDa isoform, Transcription, Expression, Regulation

1 Introduction

The Na^+/Ca^{2+} exchanger is one of the two Ca^{2+} efflux systems of eukaryotic cells. Its role is especially important in excitable tissues, e.g. in heart (see [1, 2] for recent reviews). In cardiac myocytes the action potential stimulates the opening of voltage-dependent calcium channels. The Ca^{2+} influx causes further cytoplasmic Ca^{2+} increase by inducing Ca^{2+} release from the sarcoplasmic reticulum and thus determines the contraction of myofibrils. Termination of contraction is brought about by lowering the myoplasmic Ca^{2+} by accumulation in the sarcoplasmic reticulum and by extrusion primarily by the Na^+/Ca^{2+} exchanger of the plasma membrane. The exchanger can also function in the opposite direction, but its role in Ca^{2+} influx is less clear: the reverse exchange activity could directly trigger contraction in concert with the Ca^{2+} channels.

The lack of specific inhibitors has complicated the study of the exchanger role. However, it is now known that in neurons also the Na^+/Ca^{2+} exchange activity is crucial to reduce intracellular Ca^{2+} concentration, after its increase due to its influx through Ca^{2+} channels to promote neurotransmitter release at the synapses. Na^+/Ca^{2+} exchange activity is also essential to phototransduction in the outer segment of rod and cone photoreceptors, since the Ca^{2+} decrease triggers a cascade of events determining the dark-state recovery and light adaptation. However, the primary structure of the retinal exchanger is unrelated to that of the cardiac and neuronal exchanger family [3], and its ion transport specificity and stoichiometry: 4 Na^+:1 Ca^{2+}:1 K^+ are also different [4]. The retinal exchanger will not be discussed in this chapter, the focus will be on the functional and molecular aspects of the Na^+/Ca^{2+} exchanger and its isoforms.

2 Transport Activity

The Na^+/Ca^{2+} exchange activity was first identified in cardiac atria and in squid axons by measuring ion fluxes with isotopic tracers [5, 6]. The experiments indicated a Na^+-sensitive, ouabain-insensitive Ca^{2+} transport, which was responsible for the extrusion of the latter ion from the cell. In squid axons Ca^{2+} influx was obtained by replacing external Na^+ with Li^+. The influx of Ca^{2+} was potentiated by increasing internal Na^+, conclusively indicating coupling between Ca^{2+} influx and the outward movement of Na^+. The increased Ca^{2+} influx was indirectly linked to the action of cardiac glycosides (digitalis), which inhibit the Na^+/K^+ ATP-ase, inducing a rise in intracellular Na^+ and a stimulation of Ca^{2+} influx through the reverse exchange activity.

The use of cardiac membrane vesicles has provided conclusive evidence for the reversibility and electrogenicity of the exchange process, and led to the determination of the transport stoichiometry of 3 Na^+:1 Ca^{2+} [7–9]. The generation of a membrane potential during the transport process was subsequently verified by electrophysiological measurements in cardiac myocytes [10], and the development of the "giant patch" technique has considerably advanced information on the exchanger function and regulation. A maximal turnover number of $\sim$5000 s^{-1}, and Kms for cytosolic Ca^{2+} and Na^+ of 4 μM and 18 mM, respectively, were thus estimated [11, 12]. The reaction mechanism derived from these studies predicted two consecutive transport reactions with the ion-binding sites oriented either toward the cytosol or the extracellular space. After binding the three Na^+ ions at one membrane site, the ions are translocated and released at the opposite site, in the most electrogenic step of the process. Ca^{2+} then binds at the site of Na^+ release to be transported across and released to the opposite site. A negative regulation by Na^+ appears to occur when the exchanger is fully loaded with Na^+ and a second inactive state is caused by the absence of Ca^{2+}. Its binding regulates positively both the forward and reverse modes of exchange activity, cytosolic Ca^{2+} also removing the inactivation by Na^+ [13, 14]. These regulatory aspects will be discussed further below.

3
The NCX Genes

After the cloning of the cardiac exchanger cDNA [15], the cloning of other isoforms soon followed. The deduced primary structure of the cardiac exchanger (NCX1) indicated a protein of 970 amino acids [15–17]. The human gene was located on chromosome 2 (p21–p23), close to the STS D2S2328 marker [18, 19], whereas the mouse gene was located on chromosome 17 [20]. The gene is contained in a $\sim$200 kb region and is composed of 12 exons (Fig. 1). The sequence encoding the exchanger ($\sim$3 kb) resides in a very large exon (exon 2, 1.8 kb) and in exons 3–12 [19]. Isoforms NCX2 and NCX3 are products of different genes, located on mouse chromosomes 7 and 12, and on human chromosomes 19 and 14, respectively (unpublished results) and are expressed preferentially in brain and skeletal muscle [20, 21]. They share 65 and 73% amino acid sequence identities with NCX1, respectively. The organization of the NCX3 gene is similar to that of NCX1, with a large exon 2 of 1.8 kb followed by several small exons [19]. The similarity in gene organization and sequence suggests that NCX genes could have evolved from a common ancestor that could resemble the putative precursor identified in *Caenorhabditis elegans* [19]. Other NCX homologues have been cloned from organisms phylogenetically distant, e.g. fruit flies and squids: they share 47 and 58% sequence homology with NCX1, respectively [22–24].

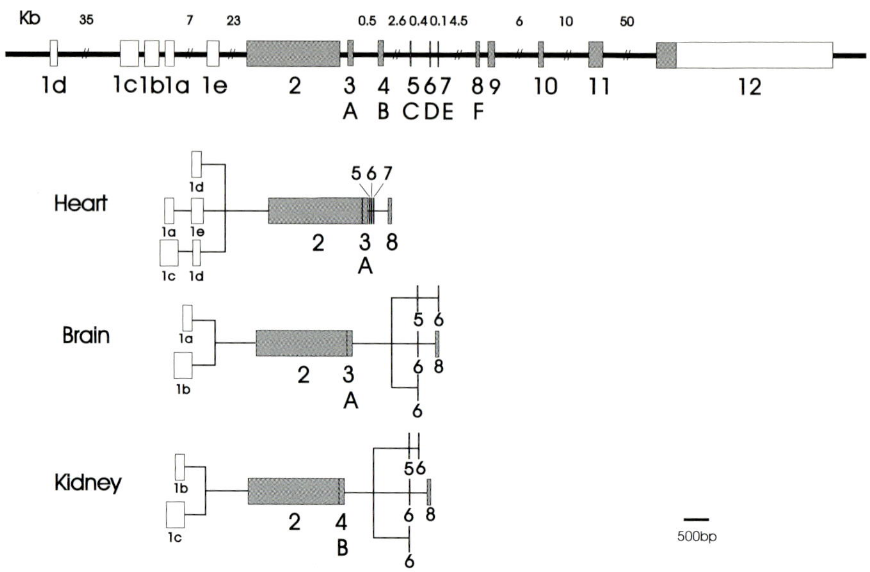

Fig. 1. The organization of the NCX1 gene. The intron-exon organization of the NCX1 gene is shown in the top panel. The exons are numbered from 1 to 12, the alternatively spliced exons 3–8 are also indicated with letters (*A–F*). The exon size is on scale, the intron length is given in kb. The coding regions are represented in grey, the untranslated exon sequences are in open boxes. The alternative exon composition of the mature NCX1 mRNAs produced in heart, brain and kidney is shown in the lower panels

3.1 Alternative Splicing

The mature NCX1 mRNA contains a 5′ untranslated leader sequence generated by tissue-specific alternative splicing of exons 1 (a–e) [25]. A large 3′ untranslated region is contained in the mature 7 kb mRNA as part of exon 12 (~3 kb).

The NCX1 gene region comprising exons 3–8 (routinely termed exons A–F) is differently assembled by alternative splicing in a tissue-specific manner: a large number of splicing variants of NCX1 have been described [26–30]. Splicing variants in the large loop region have been detected also in the NCX 3 gene [31], but so far not in the NCX2 gene.

The spliced region encodes the C-terminal segment of the large cytosolic loop protruding between transmembrane domains 5 and 6 (Fig. 2A). As shown in Fig. 1, exons A and B are mutually exclusive: exon A is typically present in the cardiac and neuronal cell transcripts, exon B is found in those of kidney. Exons C–F are often excluded in the NCX1 mRNA, but are present in the cardiac transcript, although heart also produces an isoform-lacking exon F.

Alternative splicing involving the 3′ region of the NCX1 mRNA has been discovered by expressing a 6 kb NCX1 mRNA in HEK293 cells. Three splicing variants modify the C-terminal region of the exchanger, replacing the last three hydrophobic segments with a different hydrophobic sequence. In a third isoform type the entire

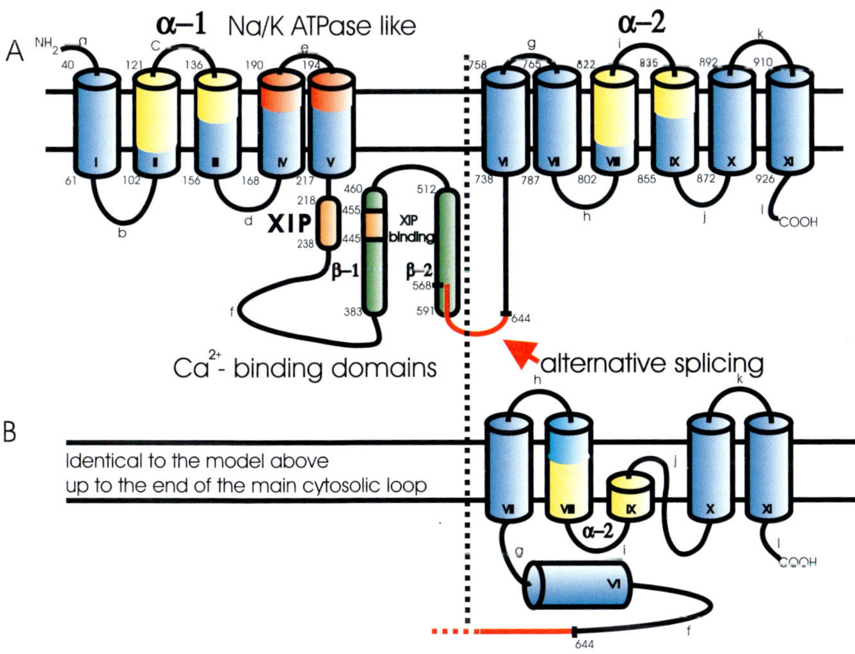

Fig. 2. Membrane organization of the exchanger, functional and regulatory domains. Following the original proposal (**A**), the exchanger is represented with 11 hydrophobic segments spanning the membrane (in *blue*). The regions which have been tentatively identified as components of the ion-translocating pathway are the α-1 and α-2 repeats (in *yellow*) and the Na$^+$/K$^+$ ATP-ase-like region (in *red*). The regulatory domains of the large intracellular loop are shown in different colours: *XIP* XIP binding, the Ca^{2+} binding domains in the β-1 and β-2 regions, and the region of alternative splicing. The numbers refer to the amino acid positions in the mature NCX1 protein. The model also applies to the NCX2 and NCX3 isoforms, but splicing variants of NCX2 have not been identified as yet. The truncated exchanger isoform encoded by exon 2 terminated after amino acid 568. The lower panel (**B**) shows the recently proposed alternative topological model of the exchanger [64] in which the portion of the exchanger following the large intracellular loop is predicted to contain only four transmembrane domains

sequence encoding the six C-terminal hydrophobic segments was deleted and a stop codon in the spliced-in exon truncated the NCX1 frame (Fig. 3d) [32, 33]. A NCX1 isoform in which the last hydrophobic segment was replaced by a different sequence was recently cloned from the BALB/c mouse heart [34], the splicing occurring at the same site which is operative in HK293 cells.

4
Control of NCX1 Transcription

The 7 kb transcript, corresponding to the full-length NCX1 mRNA, is abundantly present in heart, brain and kidney and at lower levels in several other tissues [16, 17,

3'exon 2 5'intron 2

a) GAATTCCAGAATGATGAAATTGT/*G***TAA***GTTCTATATTATATATGTGTGTGTGTGTGTG*

E F Q N D E I V •

3'exon 2 5'exon 2

b) GAATTCCAGAATGATGAAATTGT/TAG/GTTG**TGA**CAGTTGGAAGTGTCATGTACAACA

E F Q N D E I V **R** **L** •

3'exon 2 5'exon 2

c) GAATTCCAGAATGATGAAATTGT/GTTG**TGA**CAGTTGGAAGTGTCATGTACAACA

E F Q N D E I V **L** •

3'exon 2 5'exon 3 exon 12 (UTR)

d) GAATTCCAGAACGATGAAATTGT/CAAAACAATATCAGTCA/AGCTT**TAG**GGGGAAGT

E F Q N D E I V K T I S V K **L** •

Fig. 3. C-terminal sequences of the truncated NCX1 isoforms. The sequence of the human NCX1 exon-intron 2 junction is shown in (**a**). The intron sequences are in *italics* and the stop codon in intron 2 that terminates the NCX1 frame is in *bold*. The truncated isoforms encoded by the circular exon 2 transcript [35] are shown in (**b**) and (**c**): the three-base difference is related to the alternative 5′ splicing site of exon 2. The C-terminal sequence of the short isoform spliced in HK293 cells is shown in (**d**)

29]. Additional mRNA bands have also been detected, a particularly prominent one of 1.7 kb only being detected with probes from the 5′ end of the NCX1 cDNA: it corresponds to a circularized form of exon 2, which will be discussed below [35].

4.1 Regulation of the mRNA Levels

Modulation of the levels of NCX1 mRNA by various agents has been reported in several cell types, both during development and in some pathological conditions. NCX1 gene transcription is up-regulated in cardiac hypertrophy as is the transcription of several other genes [36]: the hypertrophic response to α_1-adrenergic stimulation occurs via the protein kinase C-inositol trisphosphate signalling route, Ca^{2+} evidently increasing the transcription of the NCX1 gene [37]. Positive regulation of NCX1 gene transcription by serum (also mediated by protein kinase C or by tyrosine kinases) and down-regulation by glucocorticoids have been observed in vascular smooth muscle cells [38]. In contrast to heart, the protein kinase C activator phorbol ester negatively influences NCX1 transcription in renal epithelial cells [39]. The NCX2 transcripts are rapidly down-regulated during the maturation of cultured cerebellar neurons (granule cells) at variance with those of NCX1 and of NCX3, which are not affected [39b]. The down-regulation is mediated by Ca^{2+} via the Ca^{2+}-calmodulin-dependent protein phosphatase calcineurin.

4.2 Promoter Structure and Activity

The NCX1 promoter structure has been analyzed by Scheller et al. [25] in human and rat genomes by expressing gene reporter constructs in vascular smooth muscle cells. In this system NCX1 gene transcription is controlled by a multipartite promoter located 30 kb from the first coding exon 2 that exibits the highest promoter activity. It has the high GC content characteristic of housekeeping promoters, its activity being increased by at least two additional promoter regions located 1.5 and 3.5 kb upstream [25].

Recently, it was found that NCX1 transcription initiates from different promoter regions that are alternatively operative in heart, brain and kidney [40]: this may account for the ubiquitous or inducible expression of the gene in different cell types. In heart the expression of the NCX1 gene is regulated by the proximal promoter whose minimal structure resides in the −184/−172 bp region [41]. The region includes several enhancer elements, the CArG binding, the serum responsive factor (SRF), the GATA element binding the transcription factor GATA-4 and the E box that mediates β-adrenergic up-regulation, and is a potential target for helix-loop-helix transcription factors. Promoter mutagenesis and deletion analysis have identified an additional enhancer in the first intron (+103/+134) that is also required for α-adrenergic stimulation.

The down-regulation of the NCX1 mRNA levels by protein kinase C in kidney (see above) could be due to a negative control of transcription initiation: in the kidney promoter SRE is absent, whereas several cAMP responsive elements such as CREB and ATF are present. Binding sites for tissue-specific transcription factors have also been identified in the kidney promoter. The brain NCX1 promoter is structurally different, i.e. it includes binding sites for ubiquitous transcription factors (Sp1) and has features of a housekeeping gene promoter [40].

4.3 Control of RNA Splicing

The relative intensity of the NCX1 mRNA bands indicates higher levels of the small 1.7 kb mRNA than of the full-length mRNA (7 kb). Quantitative polymerase chain reaction (PCR) analysis on different segments of the full-length cDNA has indicated that human heart contains more copies of the mRNA species that only contain the exon 2 sequence than of the full-length mRNA (unpublished observations). A fraction of the mRNA (3:1) appeared to be truncated in the 5′ sequence of intron 2.

The human intron 2 sequence was found to contain, at its 5′ end, a minisatellite with 10–16 GT repeats (N. Gabellini, to be submitted). The 5′ segment of intron 2 was inserted in the 3.7 kb NCX1 cDNA to reconstruct the exon 2/intron 2 junction. The mRNA transcribed in transfected HK 293 cells was ~2.5 kb, i.e. shorter than the regular 3.7 kb transcript. Apparently, the 5′ segment of intron 2 included a splicing enhancer. The 3′ end of the truncated transcript corresponded to the 3′ end of exon 2. The size of the truncated mRNA (1.8 kb) was increased by polyadenylation and by the presence of a short 3′ untranslated sequence of the vector.

At first glance, the truncated mRNA appeared too large to correspond to the NCX1 mRNA band of 1.7 kb since it contained at least 1.8 kb of exon 2 and the poly(A) tail as well. However, a recent study has identified the 1.7 kb mRNA species with a circularized exon 2 transcript [35]. The circular RNA could migrate faster on agarose gels explaining the apparent size discrepancy between the truncated exon 2 transcript and the 1.7 kb mRNA. Circularized transcripts have also been identified in other gene products [42], but their functional significance is unknown. The high abundance of the circular exon 2 transcript suggests a control on the NCX1 pre-mRNA splicing.

5 Exchanger Structure and Function

5.1 The Exchanger Polypeptides

As mentioned above, the deduced exchanger sequence contains 970 amino acids, including the N-terminal signal peptide (32 amino acids), corresponding to a mass of 108 kDa. However, sodium dodecyl sulfate (SDS) gel analysis and purification attempts have shown that the situation may be more complex.

The exchanger of heart sarcolemma was first proposed to be a molecule of 30–33 kDa [43, 44]. Specific antibodies raised against this polypeptide also recognized larger proteins of 70 and 140 kDa. The synaptic plasma membrane Na^+/Ca^{2+} exchanger was immunologically identified as a protein of 70 kDa: a 33 kDa protein, probably a proteolytic product, was inconsistently also present in the preparation [45]. Subsequent purification work from heart sarcolemma led to the identification of two prominent bands of 120 and 70 kDa [46], but a protein of 160–130 kDa was also recognized by antibodies directed to the 120 kDa species under non-reducing conditions. Mutagenesis work has shown that the 130–160 kDa component is an internally bonded version of the 120 kDa protein [47]. Cys 14 or Cys 20 near the N-terminus binds to Cys 792 in loop h, which in the alternative topological model of the protein (Fig. 2B), discussed in detail below, is also located externally. In kidney proximal tubules a highly active Na^+/Ca^{2+} exchanger of molecular mass 65–70 kDa was specifically located in the basolateral membrane [48].

Differential staining of the exchanger polypeptides of heart sarcolemmal membranes was obtained with two antibodies: one raised against the full-length protein that recognized all three exchanger components (130, 108 and 70 kDa), one raised against a peptide in the C-terminal half of the main cytosolic loop (amino acids 561–688) that reacted significantly only with the 108–120 kDa component. The 64–70 kDa isoform was stained only weakly by the antipeptide antibody, indicating it to be the N-terminal segment of the 108 kDa protein (Gabellini, unpublished results). It could be a proteolytic product of the larger protein, but it could also be the product of the NCX1 exon 2 transcript mentioned above: thus, the 70 kDa species could be heterogeneous, i.e. it could contain at least two components.

A 70 kDa species was recently isolated from bovine cardiac sarcolemmal vesicles [49]. In agreement with results from our laboratory, an antipeptide antibody against residues 622–644 failed to react with the 70 kDa protein, which was recognized instead by antibodies against the N-terminal portion of the exchanger. Evidently, the 70 kDa polypeptide corresponded to the N-terminal portion of the full-length protein: when reconstituted, it had transport activity, as was the case for expressed isoforms, truncated immediately after exon 2, or containing seven additional amino acids after the truncation (Fig. 3) [33]. The truncated isoform encoded by exon 2 has also been expressed in other laboratories and shown to have Ca^{2+} transport activity [35]. Thus it seems clear that the 70 kDa C-terminal peptide present in the expression products of the full-length cDNA using the vaccinia virus system [50] does not correspond to the 70 kDa species seen in heart sarcolemmal membranes.

However, even if the C-terminal hydrophobic block is apparently not essential for the transport function of the exchanger, mutagenesis studies in the transmembrane regions of homology (α-1 and α-2 repeats, see below) have shown that mutations in the conserved C-terminal portion are also disruptive [51]. One could think of the two hydrophobic blocks as the two halves of a dimer, and cautiously propose that two 70 kDa isoforms associate to form a functionally active dimer. There is some support for this proposal: e.g. the lack of significant staining of the 130–160 kDa component by the antibody against amino acids 561–688 (Gabellini, unpublished results), as well as the observation that its staining intensity was decreased by reducing agents [46], together with the appearance of a band of about 130 kDa when expressing the 67 kDa isoform [33]. A portion of the largest NCE component could thus be a dimer of the truncated 70 kDa isoform, another portion being the internally bonded variant of the 108 kDa component indicated by the Cys-mutational work [47] described above.

5.2 Cell Topology

The Na^+/Ca^{2+} exchanger has been located in the transverse (T) tubular membranes of ventricular myocytes. This location could be relevant to the reverse mode of exchange activity, since the T tubules are associated with the feet of the junctional sarcoplasmic reticulum (SR) where the Ca^{2+} release channels are concentrated [52]. Even if L-type Ca^{2+} channels are also found in this sarcolemmal domain, a small amount of Ca^{2+} entering through the exchanger could induce Ca^{2+} release from the SR: this is supported by the observation that a Na^+ current is associated with SR Ca^{2+} release [53]. The presence of the Na^+/Ca^{2+} exchanger in the T tubules may also be related to its association with the cytoskeletal protein ankyrin [54, 55].

The exchanger is also concentrated at the presynaptic membrane of nerve terminals, suggesting its involvement in the regulation of the release of neurotransmitters [56]. The concentration of the exchanger in the synaptic regions of dendrites is in line with the necessity to efficiently reduce the high Ca^{2+} concentration reached in these cell domains during excitation [57].

5.3 Post Translational Modifications

A hydrophobic N-terminal signal peptide is removed from the exchanger molecule in the endoplasmic reticulum [58]. Surprisingly, this signal sequence is not required for the correct membrane targeting of the protein: expression experiments on the cardiac, renal or brain NCX1 isoforms, with and without mutations in the signal sequence, showed that the mutatant proteins were still correctly inserted in the plasma membrane and were active [59–61]. Other structural features are thus likely to be important for correct membrane targeting: the expressed 70 kDa isoform, carrying the signal peptide but lacking the six C-terminal hydrophobic segments, was partially mis-targeted [33], suggesting that the C-terminal portion of the protein could also be important.

A second post-translational modification of the exchanger is its glycosylation. Of the six potential N-glycosylation sites, only one (Asn-9) is glycosylated. However, mutations of this site had no effect on the functional properties of the expressed protein [58].

5.4 Membrane Topology

The hydrophobicity profile of the exchanger was originally interpreted to propose 11 transmembrane spanning segments, with the N-terminus at the outer side of the membrane, a large cytoplasmic loop between the fifth and the sixth transmembrane domains followed by six transmembrane segments, and the internal C-terminus [15] (Fig. 2A). Immunofluorescence studies with antipeptide antibodies have provided support for this topological model [62, 63]. However, an alternative proposal has recently resulted from experiments on the accessibility to sulfhydryl reagents of cysteines inserted in the molecule by site-directed mutagenesis [64]. The original 11 transmembrane domain topology was confirmed up to the large cytosolic loop, but that of the C-terminal hydrophobic block was modified (Fig. 2B), since cysteines in the loop between transmembrane segments 6 and 7 were found to be accessible from the cytosol. The sixth hydrophobic segment was thus shifted to the cytosolic site, as transmembrane domain 9, which was proposed to contain a motif (GIG) similar to that recently found in K^+ channels (GYG) [64, 65].

5.5 The Path for Ions Across the Exchanger Molecule

Two regions of the cardiac Na^+/Ca^{2+} exchanger located near the extracellular surface of transmembrane spanning segments 2–3 and 8–9 of the original model [15] are highly conserved in NCX2 and 3 and also in the retinal rod $Na^+/Ca^{2+}/K^+$ exchanger (Fig. 2A). A search for homologies with the Drosophila exchanger has led to the identification of two new motifs termed α- and β-repeats [22], corresponding to two conserved regions. The α-1 repeat between transmembrane 2 and 3 is also homologous to the α-2 repeat between transmembrane segments 8 and 9 of the original topological model (Fig. 2A). The β-repeats in the large hydrophilic loop also exhibit internal homology. It thus appears likely that the two hydrophobic blocks at the N- and C-terminal sites of the large cytosolic loop may have arisen from gene duplication. In the revised topological location the α-repeats are predicted to be located on opposite membrane sites (Fig. 2B) [64]: this would be analogous to the water channel aquaporin, whereas in the original model the two duplicated domains could be orientated as in ion channels [66, 67]. Expressed exchangers with mutations in the α-repeats generally lose activity [64].

The region encompassing the extracellular loop between transmembrane segments 4 and 5 (Fig. 2A) shows similarity to the Na^+/K^+ pump (46% identity in a 24 residues stretch) and was also tested by site-directed mutagenesis [64]. Mutation in this region often abolished the transport ability of the expressed protein. Interestingly, mutations of charged residues in hydrophobic regions different from the conserved portions were mostly ineffective.

6 Regulatory Functions of the Large Cytosolic Loop

The first evidence that the exchanger activity could be regulated by cytosolic Ca^{2+} and ATP came from electrophysiological studies on squid giant axons [68]: activation of the exchanger by Ca^{2+} is a consequence of its high affinity binding to a site different from that of Ca^{2+} transport. Electrophysiological studies on giant sarcolemmal patches [11] showed that the stimulations by Ca^{2+} and ATP were distinct phenomena, the former being much more rapid and reversible. These studies also showed that the exchange current significantly decreased when Na^+ was applied to the intracellular side: this negative regulation, defined as a Na^+-dependent inactivation, has been briefly mentioned above. Moreover, ATP was shown to modulate the regulations by Ca^{2+} and Na^+. All these regulatory processes were abolished by the application of chymotrypsin at the cytosolic membrane surface; hence, they were mediated by the cytosolic segments of the exchanger. In agreement with this, the regulation was abolished in mutant proteins lacking most of the main cytosolic loop, which nevertheless maintained the basic exchange function [69, 70].

6.1 Regulatory Ca^{2+}

The regulatory Ca^{2+}-binding site was located within amino acids 371–525 in NCX1 by expressing peptides of the loop region and probing them in gel Ca^{2+} overlay experiments [71]. The NCX1 Ca^{2+}-binding site includes a cluster of negatively charged residues that is highly conserved throughout species, and also in NCX2 and NCX3 [72]. An intragenic sequence conservation was also detected in Drosophila in the region encoding the Ca^{2+}-binding sites [22]. These are the homologous regions termed β-1 and β-2 (see above), corresponding to NCX1 amino acids 383–460 and 512–591, respectively. The Ca^{2+}-binding peptide within the stretch 371–525 overlaps the β-repeats (Fig. 2A). Although the Drosophila exchanger is negatively regulated by Ca^{2+}, it still exhibits the same Ca^{2+}-binding motif [72]: a sequence different from that of the Ca^{2+}-binding site is likely to mediate the regulatory process. The regulating process has been analyzed in two spliced isoforms of the Drosophila exchanger differing only by five residues downstream of the Ca^{2+}-binding site. Their response to Ca^{2+} differed, i.e. the inhibition by Na^+ was stabilized in one isoform type and decreased in the other [73]. Thus, tissue-specific splicing in this region could confer to the exchanger regulatory properties adaptable to the requirement of Ca^{2+} homeostasis in different cell types.

6.2 Exchanger Inhibitory Peptide

A sequence composed of basic and hydrophobic residues resembling a calmodulin-binding domain at the N-terminal portion of the cytoplasmic loop (amino acids 218–238) has been shown to have an autoinhibitory function. This sequence is termed eXchanger Inhibitory Peptide (XIP) [74] and is strictly conserved in all exchanger species (see [72] for the alignment). Mutations in the XIP region resulted in the loss of the Na^+-dependent inactivation, but also perturbed the activation by Ca^{2+} [75]. Interestingly, the Ca^{2+}-binding peptide (371–525) overlaps this region. In analogy with other Ca^{2+} transporters, a peptide (amino acids 445–455 of the cardiac exchanger) cross-linked specifically to the XIP suggesting its involvement in XIP binding [76]. XIP was found to inhibit other Ca^{2+} transporters, e.g. the plasma membrane (PM) and the sarcoplasmic reticulum (SR) Ca^{2+} pumps, but not other calmodulin-binding proteins that do not transport Ca^{2+}, such as CaM kinase II [77]. A basic mechanism of regulation of these transporters could thus involve the displacement of the inhibitory peptide bound to a negatively charged region. Such a mechanism functions in the Ca^{2+} pump of the PM, where calmodulin displaces an autoinhibitory peptide. In the exchanger, the inhibitory binding of XIP could be removed by Ca^{2+} binding to the regulatory sites, making the XIP-binding site accessible to XIP (Fig. 4). The active conformation could be further stabilized by a phosphorylated peptide (see Sect. 6.3).

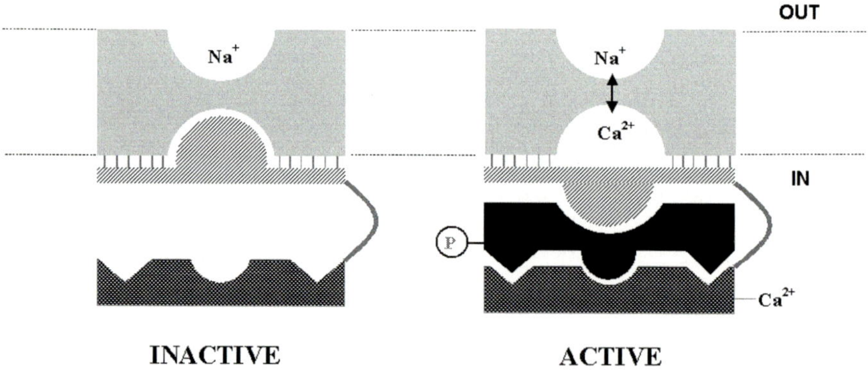

Fig. 4. Model of regulation of the Na^+/Ca^{2+} exchanger by autoinhibition and activation by Ca^{2+}. In the inactive state (*left*) the inhibitory peptide XIP (*dashed*) is visualized to occupy the ion-binding sites at the intracellular surface of the exchanger, holding back Na^+. In the active state, regulatory Ca^{2+} would bind to the intracellular Ca^{2+}-binding sites (*dotted*). This would cause the alternative binding of XIP to its specific site in the Ca^{2+}-binding peptide, a putative phosphorylated peptide (*black*) could stabilize this XIP binding. In the active conformation the ion channel in the hydrophobic core of the protein (*grey*) would be accessible to Na^+ and Ca^{2+}, that would be sequentially translocated across the plasma membrane (*dotted lines*) in opposite directions

6.3
ATP and Phosphorylation

The regulation of the exchanger by ATP is still controversial. It has been suggested that the ATP stimulation could be due to an indirect effect on the sarcolemmal lipid bilayer. Application of anionic phospholipids at the cytoplasmic surface of giant cardiac membrane patches stimulated the exchange current, much as the addition of ATP, possibly via an amino phospholipid translocase [78]. More recently, the stimulation by ATP was related to the synthesis of phosphatidyl inositol-4,5-bis-phosphate (PIP_2), since it was inhibited by the PIP-specific phospholipase C and restored by the addition of PIP_2 [79]. Antibodies against PIP_2 also reversed the ATP stimulation [80]. ATP, however, could also act by directly phosphorylating the exchanger in a kinase-mediated process or by phosphorylating a separate protein that would then mediate exchanger activation (Fig. 4). Indications for phosphorylation-linked stimulation of the cardiac exchanger were first obtained by Caroni and Carafoli [81]. In that study Ca^{2+}-calmodulin-sensitive kinases and phosphatases were proposed to be involved in the regulation. However, in other studies on membrane vesicles, ATP activation was not consistently observed, leading to the suggestion that a water-soluble protein could be involved in the activation by ATP which would be lost during membrane purification. A small cytoplasmic protein (13 kDa), isolated from squid axoplasm, was found to restore the regulation of the Na^+-dependent Ca^{2+} efflux by ATP in dialyzed squid axons [82].

Other indications for the up-regulation of the exchange activity by phosphorylation came from studies on the stimulatory effect of growth factors, such as PDGF in platelets or of insulin in heart [83, 84]. The increase in the exchange activity by Ca^{2+}-linked agonists like phenylephrine, angiotensin II and endotelin, was reversed by the selective protein kinase C inhibitor chelerythrine, whereas protein kinase C activating agents had a stimulatory effect [85]. That the activation of protein kinase C up-regulated the exchanger activity was also confirmed in rat hepatocytes, with the aid of the specific exchanger inhibitor KB-R7943 [86]. Evidence for the direct phosphorylation of the exchanger by protein kinase C was provided in transfected cells stably expressing NCX1 [87]. The phosphorylation occurred on serine residues on at least two peptides, one of which was identified with the N-terminal segment of the main cytosolic loop (amino acids 240–475): the exchanger activation correlated well with the extent of protein phosphorylation, whereas ATP depletion caused inhibition of the exchange activity and protein dephosphorylation. A comparative study of the phosphorylation-dependent regulation of the various NCX isoforms showed that the exchange activity of NCX3 was also up-regulated by protein kinase C, whereas that of NCX2 was not [88]. The phosphorylated serines were identified by site-directed mutagenesis at positions 249, 250 and 357. Two are close to the XIP; however, their substitution with alanine failed to suppress protein kinase C activation. Nevertheless, mutant proteins lacking the large cytosolic loop lost the response.

At variance with heart, the exchange activity in primary cultures of cortical neurons was sensitive to genistein, a specific tyrosine kinase inhibitor. By contrast, activation of protein kinase C had no effect [89]. In line with this, in cultured adrenal chromaffin cells the stimulatory effect of lysophosphatidic acid was potentiated by the tyrosine phosphatase inhibitor vanadate and was inhibited by genistein [90].

6.4 Alternative Splicing and Exchanger Regulation

As discussed above all the regulatory sites of the exchanger are located in the large cytosolic loop of the protein. The regulatory Ca^{2+}, XIP, the XIP binding peptide and protein kinase C phosphorylation sites are all encoded in exon 2, and are thus present in all NCX isoforms.

Alternative splicing of primary transcripts occurs frequently in NCX1 and, less frequently, in NCX3. As mentioned above, no splicing variants of NCX2 have so far been described. The functional significance of the tissue-specific alternative splicing in the loop region has not been fully elucidated. However, phosphorylation sites could be specifically modified by this mechanism, i.e. the tissue-specific expression of isoforms with variable loop structures could explain differences in kinase sensitivity. This is supported by the observation that glia and neuronal cells express different NCX1 isoforms, containing the mutually exclusive exons B or A. When expressed in *Xenopus* oocytes the isoforms had normal activity; however, the activity of the neuronal isoform containing the product of exon A was increased by protein kinase A, even if direct evidence for phosphorylation of the exon A product could not be obtained [91]. Alternative splicing is also involved in a potential regulatory mechanism recently discovered in frog heart: a short exon of 27 bp, present and abundant only in frog cardiac transcripts, creates a Walker B (P-loop) ATP-binding site in the alternatively spliced exon of the loop region, i.e. downstream with respect to most other sites involved in the regulation of the exchanger. The frog isoform expressed in oocytes showed a sensitivity to cAMP typical of the native frog heart exchanger, which was absent in a deletion mutant lacking the nine amino acid segment [92].

7 Conclusions

One of the problems that hampers the study of the Na^+/Ca^{2+} exchanger is the lack of a simple enzymatic test to assess its activity. Membrane incorporation of (semi)purified preparations and expression of recombinant variants, accompanied by measurements of ion fluxes, are necessary. Despite these difficulties, significant progress is being made, particularly in the area of regulation, where a multiplicity of mechanisms ranging from allosteric ion effects to protein phosphorylation has been described. A problem which needs to be explored and clarified is the reason for the existence of numerous isoforms; very likely, their existence responds to different demands of various cell types (or even cell domains in complex systems, e.g. neurons) in terms of Ca^{2+} homeostasis. The definition of the differential functional properties of the alternatively spliced isoforms is thus an urgent necessity.

One area where substantial progress is being made is that of mutational analysis: the work is providing encouraging results on the matter of the membrane topology of the exchanger and on the path for the translocation of Ca^{2+} and Na^+ through the protein.

References

1. Reeves JP (1998) J Bioenerg Biomembr 30:151
2. Philipson KD (1999) Sodium-calcium exchange. In: Carafoli E, Klee C (eds) Calcium as a cellular regulator. Oxford University Press, New York, Oxford, chap 11, p 279
3. Reiländer H, Achilles A, Friedel U, Maul G, Lottspeich F, Cook NJ (1992) EMBO J 11:1689
4. McNaughton PA (1995) Cell Calcium 18:275
5. Reuter H, Seitz N (1968) J Physiol 195:451
6. Baker PF, Blaustein MP, Hodgkin AL, Steinhardt RA (1969) J Physiol 200:431
7. Pitts BJR (1979) J Biol Chem 254:6232
8. Caroni P, Reinlib L, Carafoli E (1980) Proc Natl Acad Sci USA 77:6354
9. Reeves JP, Hale CC (1984) J Biol Chem 259:7733
10. Kimura J, Miyamae S, Noma A (1987) J Physiol London 384:199
11. Hilgemann DW (1990) Nature 344:242
12. Hilgemann DW, Nicoll DA, Philipson KD (1991) Nature 352:715
13. Hilgemann DW, Matsuoka S, Nagel GA, Collins A (1992) J Gen Physiol 100:905
14. Hilgemann DW (1996) Biophys J 71:759
15. Nicoll DA, Longoni S, Philipson KD (1990) Science 250:562
16. Kofuji P, Hadley RW, Kieval RS, Lederer WJ, Schulze DH (1992) Am J Physiol 263:C1241
17. Komuro I, Wenninger KE, Philipson KD, Izumo S (1992) Proc Natl Acad Sci USA 89:4769
18. Shieh BH, Xia Y, Sparkes RS, Klisak I, Lusis AJ, Nicoll DA, Philipson KD (1992) Genomics 12:616
19. Kraev A, Chumakov I, Carafoli E (1996) Genomics 37:105
20. Nicoll DA, Quednau BD, Qui Z, Xia YR, Lusis AJ, Philipson KD (1996) J Biol Chem 271:24914
21. Li Z, Matsuoka S, Hryshko LV, Nicoll DA, Bersohn MM, Burke EP, Lifton RP, Philipson KD (1994) J Biol Chem 269:17434
22. Schwarz EM, Benzer S (1997) Proc Natl Acad Sci USA 94:10249
23. Ruknudin A, Valdivia C, Kofuji P, Lederer WJ, Schulze DH (1997) Am J Physiol 273:C-257
24. He Z, Tong Q, Quednau BD, Philipson KD, Hilgemann DW (1998) J Gen Physiol 111:857
25. Scheller T, Kraev A, Skinner S, Carafoli E (1998) J Biol Chem 273:7643
26. Nakasaki Y, Iwamoto T, Hanada H, Imagawa T, Shigekawa M (1993) J Biochem 114:528
27. Kofuji P, Lederer WJ, Schulze DH (1993) Am J Physiol 265:F598
28. Kofuji P, Lederer WJ, Schulze DH (1994) J Biol Chem 269:5145
29. Lee SL, Yu ASL, Lytton J (1994) J Biol Chem 269:14849
30. Furman I, Cook O, Kasir J, Rahamimoff H (1993) FEBS Lett 319:105
31. Quednau BD, Nicoll DA, Philipson KD (1997) Am J Physiol 272:C1250
32. Gabellini N, Iwata T, Carafoli E (1995) J Biol Chem 270:6917
33. Gabellini N, Zatti A, Rispoli G, Navangione A, Carafoli E (1996) Eur J Biochem 239:897
34. Shi S, Chang B, Brunnert SR (1998) Biochem Genet 36:119
35. Li XF, Lytton J (1999) J Biol Chem 274:8153
36. Moalic JM, Charlemagne D, Mansier P, Chevalier B, Swynghedauw MD (1993) Circulation 87:IV-21
37. Reinecke H, Vetter R, Drexler H (1997) Cardiovascular Res 36:216
38. Smith L, Smith JB (1994) J Biol Chem 269:27527
39. a) Smith L, Porzig H, Lee HW, Smith JB (1995) Am J Physiol 269:C457
    b) Guerini D, Lei L, Genazzani A, Carafoli E (2000) J Biol Chem, in press
40. Nicholas SB, Yang W, Lee SL, Zhu H, Philipson KD, Lytton J (1998) Am J Physiol 274:H217
41. Cheng G, Hagen TP, Dowson ML, Barnes KV, Menick DR (1999) J Biol Chem 274:12819
42. Zaphiropoulos PG (1997) Mol Cell Biol 17:2985
43. Soldati L, Longoni S, Carafoli E (1985) J Biol Chem 260:13321
44. Longoni S, Carafoli E (1987) Biochem Biophys Res Commun 145:1059
45. Barzilai A, Spanier R, Rahamimoff H (1987) J Biol Chem 262:10315
46. Philipson KD, Longoni SL, Ward R (1988) Biochim Biophys Acta 945:298
47. Santacruz-Toloza L, Ottolia M, Nicoll DA, Philipson KD (2000) J Biol Chem 275:182
48. Dominguez JH, Juhaszova M, Kleiboeker SB, Hale CC, Feister HA (1992) Am J Physiol 263:F945
49. Saba RI, Bollen A, Herchuelz A (1999) Biochem J 338:139
50. Iwata T, Galli C, Dainese P, Guerini D, Carafoli E (1995) Cell Calcium 17:263
51. Nicoll DA, Hryshko LV, Matsuoka S, Frank JS, Philipson KD (1996) J Biol Chem 271:13385

52. Frank JS, Mottino G, Reid D, Molday RS, Philipson KD (1992) J Cell Biol 117:337
53. Leblanc N, Hume JR (1990) Science 248:273
54. Li Z, Burke EP, Frank JS, Bennet V, Philipson KD (1993) J Biol Chem 268:11489
55. Chen F, Mottino G, Shin VY, Frank JS (1997) J Mol Cell Cardiol 29:2621
56. Luther PW, Yip RK, Bloch RJ, Ambesi A, Lindenmayer GE, Blaustein MP (1992) J Neurosci 12:4898
57. Reuter H, Porzig H (1995) Neuron 15:1077
58. Hryschko LV, Nicoll DA, Weiss JN, Philipson KD (1993) Biochem Biophys Acta 1151:35
59. Toth MS, Nicoll DA, Frank JS, Philipson KD, Friedlander M (1995) Biochem Biophys Res Commun 212:968
60. Furman I, Cook O, Kasir J, Low W, Rahamimoff H (1995) J Biol Chem 270:19120
61. Loo TW, Ho C, Clarke DM (1995) J Biol Chem 270:19345
62. Porzig H, Li Z, Nicoll DA, Philipson KD (1993) Am J Physiol 265:C748
63. Cook O, Low W, Rahamimoff H (1998) 1371:40
64. Nicoll DA, Ottolia M, Lu L, Lu Y, Philipson KD (1999) J Biol Chem 274:910
65. Doyle DA, Cabral JM, Pfuetzner RA, Kuo A, Gulbis JM, Cohen SL, Chait BT, MacKinnon R (1998) Science 280:69
66. Agre P, Bonhivers M, Borgnia MJ (1998) J Biol Chem 273:14659
67. Catterall WA (1988) Science 242:50
68. DiPolo R (1979) J Gen Physiol 73:91
69. Matsuoka S, Nicoll DA, Reilly RF, Hilgemann DW, Philipson KD (1993) Proc Natl Acad Sci USA 90:3870
70. Condrescu M, Gardner JP, Chernaya G, Aceto JF, Kroupis C, Reeves JP (1995) J Biol Chem 270:9137
71. Levitsky DO, Nicoll DA, Philipson KD (1994) J Biol Chem 269:22847
72. Dyck C, Maxwell K, Buchko J, Trac M, Omelchenko A, Hnatowich M, Hryshko LV (1998) 273:12981
73. Omelchenko A, Dick C, Hnatowich M, Buchko J, Nicoll DA, Philipson KD, Hryshko LV (1998) J Gen Physiol 111:691
74. Li Z, Nicoll DA, Collins A, Hilgemann DW, Filoteo AG, Penniston JT, Weiss JN, Tomich JM, Philipson KD (1991) J Biol Chem 266:1014
75. Matsuoka S, Nicoll DA, He Z, Philipson KD (1997) J Gen Physiol 109:273
76. Hale CC, Blirer S, Quinn TP, Peletskaya EN (1997) Biochem Biophys Res Commun 236:113
77. Enyedi A, Penniston JT (1993) J Biol Chem 268:17120
78. Hilgemann DW, Collins A (1992) J Physiol 454:59
79. Hilgemann DW, Ball R (1996) Science 273:957
80. Huang CL, Feng S, Hilgemann DW (1998) Nature 391:803
81. Caroni P, Carafoli E (1983) Eur J Biochem 132:451
82. Di Polo R, Berberian G, Delgado D, Rojas H, Beauge L (1997) FEBS Lett 401:6
83. Cirillo M, Canessa ML (1991) Ann NY Acad Sci 639:169
84. Ballard C, Mozaffari M, Schaffer S (1994) Mol Cell Biochem 135:113
85. Ballard C, Schaffer S (1996) J Mol Cell Cardiol 28:11
86. Ikari A, Sakai H, Takeguchi N (1998) Eur J Pharmacol 360:91
87. Iwamoto T, Pan Y, Wakabayashi S, Imagawa T, Yamanaka HI, Shigekawa M (1996) J Biol Chem 271:13609
88. Iwamoto T, Pan Y, Nakamura TY, Wakabayashi S, Shigekawa M (1998) Biochem 37:17230
89. Wang C, Davis N, Colvin RA (1997) Biochem Biophys Res Commun 233:86
90. Tokumura A, Okuno M, Fukuzawa K, Houchi H, Tsuchiya K, Oka M (1998) Biochim Biophys Acta 1389:67
91. He S, Ruknudin A, Bambrick LL, Lederer WJ, Schulze DH (1998) J Neurosci 18:4833
92. Shuba YM, Iwata T, Naidenov VG, Oz M, Sandberg K, Kraev A, Carafoli E, Morad M (1998) J Biol Chem 273:18819

Printing: Mercedes-Druck, Berlin
Binding: Stürtz AG, Würzburg